GW01417759

BUSTING ANTI-VAX MYTHS!

SERIOUSLY EXPERT ARGUMENTS FOR THE COVID-DENIERS IN YOUR LIFE

BY

PROF. OISÍN MACAMADÁIN

(EXPERT)

First published in 2022.

This book is Copyright © Prof. Oisín MacAmadáin

For Dr. B.

Praise for
Prof. Oisín MacAmadáin!

'At last! A book to *really* 'piss off' the anti-vaxxers. Je l'adore!'

President Macaroni

'And Prof. MacAmadáin was never a WEF young leader? It's scarcely credible – here is someone who truly 'gets it'!'

Santa Klaus

'Wait, am I in this book? Let me make sure I get my hair right.'

President Trudy-Wudy

'Phew, it was getting a bit worrying for a second there.... thank God for Oisín.'

CEO of Pfizzle

'The cream of Irish society has been treated to my colleague Oisín's wisdom in the pages of *The Oirish Times* nearly every day for the last two years. That he should now set his thoughts down into a book is simply the icing on the cake. We Irish truly are the best at everything and our response to the pandemic really demonstrates this. Oisín's book encapsulates this truth marvellously.'

Gubnet O'Foole, *Oirish Times* correspondent in residence.

'Extremely smart, erudite, intelligent. A true polymath...the expert of experts. What would we do without him?'

The author

FOREWORD BY DR. ANTHONY FAUCET

I'll never forget the first time I met Prof. Oisín MacAmadáin. Not only did his delightful Irish brogue charm the pants off me, but I knew then and there that this was a man whose expertise I would one day need to call upon.

However, I must admit that when I received an invitation to the opening of The Termonfeckin Institute of Expertise (T.I.E.), I was unsure as to whether to accept. Much to my shame, I had not actually heard of the bustling Irish metropolis of Termonfeckin in beautiful Co. Louth but when I read more about the vision that Prof. Oisín MacAmadáin had for the place, well, I knew I just had to meet him.

The Termonfeckin Institute of Expertise is, literally, outstanding in its own field and what a field it is too. The beautiful sights of the river Boyne in the distance and cattle munching on buttercups just opposite – you truly could not have an environment more conducive to education. Furthermore, if you were to believe its website (and I see no reason why you shouldn't), the T.I.E. has quickly risen to become one of the leading educational institutes in the world. This is all the more remarkable given that it has a faculty of just one. Yes, you guessed it: Prof. Oisín MacAmadáin himself, Provost, Head of Department and Lecturer, a real Trinity of wisdom and education if ever there was one.

And his subject is simply expertise itself. Whatever he turns his hand to, he grasps it in ways no one else can. He's far too modest to admit it himself but the man is clearly a total genius.

And that's why when the world was hit by the greatest threat it has ever faced in the form of Covid-19, I knew that this was the man I needed to call. Unsurprisingly, Oisín was his obliging self. He vowed that he'd get to the bottom of everything to do with the virus and share all his learnings with me (though I told him not to worry too much about whether it came from a bat or a cat or a pangolin or, y'know, a lab or something like that, we'd look into that part).

I was then so impressed by everything Oisín told me that I immediately made him a special advisor to the US government's taskforce for overseeing the pandemic. I think, when you read this book, you will be able to see clearly how Oisín's sharp thinking has influenced the response in the US and, ultimately therefore, most of the world. Truly, we have a lot to be grateful for when it comes to experts and none more so than Prof. MacAmadáin. Hats off to you, mi chairde mio.

But this particular book is not so much focussed on the sterling anti-Covid health policies that the likes of Prof. MacAmadáin have devised for us all. Rather, in this book, he does us all the great service of dispelling the dangerous anti-vax myths that are circulating so nefariously online. Read it to your friends, read it to your loved ones and read it to yourself when you are in the waiting room for your 7th booster. It is a book to savour: I can only commend it to you wholeheartedly.

Yours sincerely,

Dr. Anthony Faucet

INTRODUCTION

My name is Prof. Oisín MacAmadáin and I am an expert.

I have written this book to counteract the manifest untruths that are being propagated by conspiracy theorists everywhere about the greatest crisis that our world has ever faced.

This misinformation is being spread by extremists in our midst. In particular, it is being spread by conspiracy theorists, far-right activists, activists who are even further to the right of those activists, and activists who are so very right-wing that they goosestep in their sleep. These people claim to care for ideals such as democracy, freedom of expression and open scientific debate. Well, Hitler said he cared about those kinds of thing too. Or at least I think he did. I wasn't really paying attention back in school for that lesson but the point is that these people are dangerous and MUST be silenced. On that I'm sure we can all agree.

So that's why I've written this book. I have taken my many years of expertise and all of the critical thinking skills I have developed over my long and incredibly distinguished career and put down bulletproof arguments against the kinds of outlandish claims that Covid-deniers come up with. The aim is to help you shut up these lunatics once and for all. Then, we can all get back to STAYING SAFE, vaccinating EVERYONE (both human and non-human), vaccinating everyone AGAIN (many times, in fact) and doubling down on our efforts to eliminate this horrible disease FOREVER.

To paraphrase someone who would undoubtedly have been on the right side of this battle (for he would recognise that this current crisis is far worse than the one he faced, and God forgive me but I can't believe that I'm using a Brit here to make an example, but here it goes):

> 'We shall go on to the end, we shall fight this virus wherever it may be and in all its thousands of variations, now and in the centuries to come. We shall fight it in the seas and oceans (for

you never know where it might pop up next). We shall fight it with growing confidence in the air (or perhaps just stop all air travel, except for the very rich and for politicians, of course, but they'll have to wear masks, at least when being photographed), we shall defend our island, whatever the cost may be, we shall fight it on the beaches (well you can't go to the beach unless it is within 5KM of your home, that is), we shall fight it in the fields (but only on your daily exercise and not at other times) and in the streets (what on earth would you be doing on the streets?!); we shall never surrender, and even if, which I do not for a moment believe, this Island were subjugated and starving, then the rest of our comrades in governments across the world would carry on the struggle, until, the New World Order, with all its power and might, steps forth to the rescue and liberation of the old.'

I'm sure Winston would approve of my channelling of his words here.

I would like to take this opportunity to thank my editor, Máire Ní Fheadair, the Termonfeckin Institute of Expertise's sole and perennial graduate student, for casting her own (fairly) expert eyes over the text. Of course, we can't all get everything right and so any mistakes left in this book are solely her responsibility.

I would also like to dedicate this book to my father, a man who truly was no fool, and who spotted my own intellectual acumen at a young age. How sad it makes me that I can't tell you in person, Dad, that only yesterday I received a call from a top Social Media company asking me to head up their Factchecking Department… oh, if only you could be here with me now… well, I'll tell you on the phone later and, sure, you never know, maybe they'll let you out sooner than we think. In any event, every day I try my best to do justice to the MacAmadáin DNA.

Finally, thank you to my darling wife. How often we have comforted each other over the last few years and what difficult times we have faced. But even when we thought there were no more series on Netflix to watch, sure, didn't we always find just one more? All my love to you, my darling, darling Assumpta.

With all of the above said, I hope this book will help you all to arm yourselves in the great fight against misinformation that is being spread among the nether regions of the internet!

Yours sincerely and please, for the love of God, stay extremely safe,

Prof. Oisín MacAmadáin

For media queries, please write to:
<oisinmacamadain@icloud.com>

CONTENTS

Foreword by Dr. Anthony Faucet.. i

Introduction iii

Chapter One: Busting Covid-Denying Myths! 1

    Lessons from China 1

    The Declaration of Great Baloney 3

    Yes, Variants *are* Scary! 7

    A Q&A with Oisín... 10

    The Origins of Covid 15

Chapter Two: The Many Joys & Blessings of Lockdown 16

    Masks, Glorious Masks 16

    A Short Word on Dating... 19

    No, the Pandemic is NOT Leading to
    Mental Health Problems in Children 20

    The Ideal Covid Classroom: A Case Study 22

Chapter Three: Oisín's Guides to 25

    ...Factchecking 25

    ...Covering Covid in the Media 29

Chapter Four: The Lockdown Hall of Fame................................. 35

    Ireland: A Model Case Study –
    Probably the Best Lockdown in the World! 35

    Australia (hmmm might just have done the whole lockdown thing a
    bit better than us actually).. 40

    Canada (again, a country that makes me feel a little awkward
    about our own Covid performance) 44

    Austria (ok, I have to admit that this lot really did it best)............... 46

Chapter Five: The Lockdown Hall of Shame.................................50

Sweden (or 'The Sad Story of How a Liberal Utopia
Became a Far-Right Nightmare')... 50

Belarus (or 'The Land Where They Believe Vodka Kills Covid') 53

Mexico (or 'Hasta Manaña Señor Covid'?)............................... 55

Romania (or 'Oisín's Sincere and Heartfelt Advice
to the Romanian Government')... 58

Florida (or 'The Tale of Oisín's Nightmare Holiday')..................... 60

Chapter Six: Roll Up Your Sleeve Everyone!.................................65

Oh, But It is Not a Vaccine!... 67

Mass Vaccination in the Middle of a Pandemic:
Probably the Best Idea in the World...................................... 68

Getting the Vaccines to Those Who Need Them Most 69

Vaccinating Our Furry Friends 74

Chapter Seven: Enter the Anti-Vaxxers!....................................77

Robert Malone: The Greatest Antivaxxer of Them All..................... 77

Anti-Vaxxer Ringleader No 2: Peter McCollough........................... 79

Vaccine Hesitancy Disease 81

The Anti-Vaxxer 'Freedom Fighters' 84

Oisín Heads to the Far North: Meeting the Far-Right, Anti-Vax,
Canuck Truckers... 86

Conclusion: The Unvaccinated are SELFISH!.............................. 90

Chapter Eight: Busting Anti-Vax Myths!....................................93

A Dangerous Gene Therapy? 93

VAERS Reported Deaths and Adverse Events:
A Lot of Hoo-Hah About Nothing... 96

No, the Vaccines Will Not Give You a Heart Attack!...................... 99

No, the Vaccines Will Not Affect Your Fertility! (if anything they'll just
make superhero babies) 102

Chapter Nine: Quack Covid Cures 105

Ivermectin (because Covid obviously gives you worms! And turns you into a horse.... Jeez, the intellectual level of these people!) 105

The Vitamin D Debacle! 109

No One Will Take Away My Right to Eat Ice-Cream! 112

Chapter Ten: The Great Reset
(or 'The Much Needed Plan to Save Humanity from Itself') 115

Klaus Schwab: A Wise Swami for Our Times.................................. 115

The Great Termonfeckin Ejaculation 119

Oisín's Dream of the Future: The World in 2030............................ 124

CHAPTER ONE: BUSTING COVID-DENYING MYTHS!

Well, to kick this book off, we'll be tackling some of most egregious lies that Covid-deniers spread about the virus, you know, things like its supposedly dubious origins, its 'not so bad' infection fatality rate (!) or the notion that lockdowns aren't really the best idea after all ....but before we get into that myth-busting, let's start this chapter on a more positive note and first remind ourselves of the early days of the pandemic when world history was changed forever and for the better....

Lessons from China

We all have a lot to be grateful for when it comes to the Chinese. For it was they who alerted us as to the deadly seriousness of this virus in the first place.

I'll never forget watching the videos coming out of China in the early days, one with a fellow walking through a crowded square, sneezing and then him and everyone around him dropping dead. Or that video of a little village just outside Wuhan which showed hundreds of corpses being picked on by ravens in the mist and medical personnel in Hazmat suits desperately searching, and failing, to find just one person who might still be alive.

And not only did they show us just how serious this disease is but they showed us exactly the best way of dealing with it. Indeed, sometimes in science you need brave, new thinking in order to debunk old ways of seeing things...you know, real Da Vinci moments. And that's exactly what China achieved when they instigated a curfew in Wuhan, encircled it with tanks and the army, and shot anyone who dared leave their homes. Why we didn't always deal with respiratory viruses in this way, I'll never know, but whichever fella came up with this needs a Nobel Prize, no doubt about it.

And so the world looked, it listened and then it copied and, frankly, we've never looked back. In this way, scientific thought was elevated to previously unimaginable heights thanks to Chinese genius.

But it was, I have to say, by no means certain that the Chinese approach would come to be adopted everywhere. A key moment was when Covid reached Italy…. sure, how would that ultra-sociable lot go about it all? Well, they looked over at the top-drawer stuff that was happening in Wuhan and decided, as I think the old saying goes, to take the gift out of the horse's mouth: 'All citizens must now stay indoors and from this point on only perform their arias and operettas from their balconies!' sang the health minister. For me personally, this was a special moment in the history of science for it made it all the likelier that the rest of Europe would then demonstrate a bit of the old cop on and follow China's lockdown strategy (although if I could be critical about one thing and truly I hate to be but, in hindsight, the Italian government clearly did not appreciate the distance that Covid-infused droplets are capable of traversing once expelled by the well-trained lungs of a soprano or baritone….we can only forgive them, even Virgil nods or whatever it is they say, and we can but hope that next time people will be asked to sing indoors and only when they are alone in the room of course).

Ok, so Italy got on board and all eyes then turned to the UK and to a certain Prof. Neil Ferguson of Imperial College London and special advisor to the government. An expert modeller, he put his trained mind to the question of just how many people Covid would kill in the U.K and, in short, his conclusion was that it would be rather an awful lot. And so, in order to prevent this disastrous scenario, he knew a lockdown would be needed. But how to persuade the British public? Well, if the Italian government could get away with it then why couldn't Johnson and Co.? And so he advised Boris to get with the script (indeed the man had up until that point been Churchilling away on the need to keep calm and carry on or some such nonsense). Thankfully, for once Boris had the ability to recognize intelligence superior to his own and so the UK locked down nice and hard.

And, I'm glad to say, didn't we also in The Termonfeckin Institute of Expertise not make our own contribution to the clear need for lockdowns? Now, our models were even more cautious than those of Prof. Ferguson (brilliant as the man is, of course) and I'd invite you all to read our paper which is easily findable online (its title is 'Ireland's Next Great Famine: Preparing for the Impending Covid Catastrophe'). But the gist of it is that we predicted that whole suburbs of Cork, Galway and Dublin would be wiped out and that a lockdown was completely and utterly essential. Thankfully, the government listened and locked down and on Paddy's Day no less. Indeed, we at the T.I.E. remain proud of our role in Ireland's Covid response. (It would be remiss however not to note that we still strongly believe that the government should also have followed our advice for the immediate and wholescale transfer of urban populations to socially distanced camps in the Midlands but, sure, we will make sure to push for that the next time.)

In any event, whether it was thanks to China, Italy, Ferguson or indeed the work of yours truly, everywhere else pretty much then followed suit from Peru to Morocco to Germany to Turkmenistan. Indeed, people surely do know a good thing if you slap them hard enough in the face with it... and luckily enough most people knew that lockdown was a good thing and that it would save countless lives. Phew, we escaped a close one there alright!

The Declaration of Great Baloney

And I really do mean that we escaped a close one because, bizarrely, not every scientist seems to agree that lockdowns have saved countless lives and sure what if that load of eejits has been running the show...not exactly the brightest bunch, are they? I mean lockdowns equal less cases which equals less deaths....so obvious, even a toddler could grasp it. But this is a book about counteracting misinformation and so...let's get down to some of its very first myth-busting!

Now, one of the ways Covid-deniers like to add respectability to their views is to suggest that prestigious experts regard lockdowns as a big mistake. Oh, but what about The Great Barrington Declaration, they cry, referring to a little get together of some armchair and totally fringe 'scientists' who, as far as I can tell anyway, think that the best approach to Covid is to purposively encourage outbreaks of the disease in nursing homes worldwide. Well, they wouldn't say that, of course, but then what else would you expect of them?

So what is the Great Barrington Declaration all about then? There are three main protagonists to blame and they are: Prof. Jay Bhattacharya (Professor of Medicine, Stanford University), Prof. Sunetra Gupta (Professor of Theoretical Epidemiology, Oxford) and Prof. Martin Kulldorff (Professor of Medicine, Harvard until 2021). Now, you might well be thinking 'Oh but Oisín, these people sound awfully distinguished, shouldn't we be listening to what they have to say?' Honestly, folks, think harder for, as always, the devil is in the detail. For example, I very much doubt that the Stanford, Oxford or Harvard in question are the institutions we would normally associate with those names. Indeed, how do we know that they aren't just small struggling colleges in the suburbs of those towns who've decided to latch onto the names of their prestigious educational neighbours in the hope of luring in some unsuspecting Chinese students? You know, a bit like when you fly Ryanair and you think you are going to a major city only to find once you've landed that you are actually in some backwater two hours away from where you thought you were going. Prof. Bhattacharya won't spell it out but I bet you anything he is actually on the faculty of Stanford Polytechnic.

So what kind of far-right ideas are this lot putting out there, then? Let's pick them apart!

Basically, they talk about a so-called 'focussed protection' model…. protect those vulnerable to the disease, the elderly and comorbid, and otherwise allow life to continue as normal for the rest of us. Dear God, is this what passes for scientific thinking these days?! The whole thing sounds inherently discriminatory and ageist to me. Oh, so lockdown all our grannies while everyone else gets to go about as normal, is it? We live in societies where EQUALITY matters and if we let a viral pandemic destroy the principles we hold dear, then what would that say of us? If granny stays at home, then so do we all, and that's that.

And how do we even know that their strategy makes our grannies any safer? These fringe scientists no doubt would use arguments like: 'Well, if granny stays at home, then it doesn't really matter if the rest of the population gets the virus, does it? It's not like the virus can move through the walls of her house....' Oh dear Lord! How do these idiots know whether or not Covid-19 can move through walls? If it can travel all over the world from China, then God only knows what it is capable of. Honestly, the intellectual level of these people.

It's really best that we lockdown everyone, both the old and the young. That's the safest way of doing things and, to be honest, some grannies might otherwise get notions and invite their grandchildren over for tea or something...you know what they are like. So it's best they know for certain that they should keep SAFE by staying home alone. Many of them are senile anyway and need to be told what to do. Wasn't it Gandhi who said you can tell how civilised a society is from how it treats its animals? Well, the same holds true for the elderly, I would say, and keeping them confined to their homes is clearly in their best interest.

Not that they always realise this, of course. I was in a taxi the other day and the taximan gave me some sob story or other about how the government shouldn't take away his 90 year old mother's right to choose to meet with her family. It was her risk to take and she wanted to see them and so on. Honestly, what kind of woman is this?! I mean, I know for a fact that if I were 90 years old, the last thing I would want is to meet with my family. It would be nice to have a break from them at last and I'd be very grateful to the government for facilitating such an opportunity.

Of course, I wish to make abundantly clear that just because I've been talking about the mortal danger Covid poses to the elderly does NOT mean that I entertain the anti-vax idea that some people are in more danger than others. No, let me be very clear: we are all very much in danger of dying from this terrible disease, from the youngest to the oldest. Unlike this Great Barrington lot, Covid does not discriminate.

Truly, whether you are old or young, there is no 'no risk' group and so, for the love of God, please just stay indoors. Indeed, if we can conclude anything at all about the Great Barrington group, it is that it is surely the epitome of the 'let it RIP!' brigade. And what does 'RIP' stand for? Yes, you got it: 'Rest in Peace'. Honestly, the millions who would have died had this lot been in charge….well, it's too horrific even to contemplate.

Anyway, despite all that I have just said, I must admit that I was still somewhat concerned when I first learned of the emergence of this group and so I immediately wrote to my mate Dr. Faucet:

> 'From: Prof. Oisín MacAmadáin (<termonfeckineinstein@termonfeckininstitute.ie>)
>
> To: Dr. Antony Faucet
>
> Subject: VERY FRINGE GROUP OF SELF-PROCLAIMED SCIENTISTS, DANGER!!!
>
> Hey Tony!
>
> I'm extremely concerned to see the emergence of the 'Great Barrington' declaration….they could undermine all our efforts! They say we should protect granny and the vulnerable but let everyone else be free – God, if people could grasp that kind of nuanced thinking, we'd be done for! Do what you have to do! Publish a devastating takedown (I can write it if you like), call in the army, tell the people who control Joe to tell him to say that this is all the work of terrorists!
>
> Far worse is that this misinformation comes across as so convincing that even Nobel Prize winner Michael Levitt has added his signature! So I suggest we also put together a devastating takedown and factcheck of whatever work he did to win the Nobel Prize so as to make him look like a total eejit as well.
>
> I anxiously await your thoughts,
>
> Oisín'

Anyhow, I then went to read the Great Barrington website in more detail but to be honest it was one long yawn fest and my eyes glazed over. So, thankfully, I'm not so worried anymore as I don't think people these days have the attention span for reading that kind of nit-picking carryon. It's just as well as apparently some new study from Stanford (presumably the real one this time but who knows) showed that a roughly equal number of distinguished scientists support the 'focused protection' model as the lockdown one but that the latter group has more social media reach. It never ceases to amaze me how fringe, far-right ideas can end up misleading researchers into adopting positions that disagree with the science. But, then again, we truly do live in a post-truth era.

But all this talk of a Great Barrington declaration has made me wonder about whether I should do my own big pronouncement on the clear need for lockdowns and things of that general nature. I might call it 'The Termonfeckin Agreement'...no, I want something that sounds grander than that and I like the way they used 'Great', the self-aggrandizing nobs that they are.... how about 'The Great Termonfeckin Testimony'....hmmm, that doesn't make sense but it makes me think of something more biblical like.... 'The Great Termonfeckin Testament'.... yes, that's certainly better! But I wonder, I wonder.... ah, yes, perfect! And both biblical and suitably lofty in its language: 'The Great Termonfeckin Ejaculation'. I'll set about organising it straight away and I'll be sure to inform you of its findings.

Yes, Variants *are* Scary!

Well, while we can all surely look forward to my great ejaculation, we need to move on to the next myth we have to counteract. Indeed, having looked at how the stellar Chinese approach to combatting this virus was rightly adopted by nearly everywhere in the world, despite the moanings of a few right-wing scientists, let's now turn to consider the virus in more detail. There are so many things one could talk about but let's home in on, in particular, its sneaky capacity to mutate into an untold number of variants, something which only goes to highlight its manifestly evident dangerousness (although, bizarrely the Covid-deniers don't see it this way, but, then again, that's why I'm here!).

Indeed, the other day I was listening to the radio and the presenter was complaining on and on about how long the restrictions were still likely to go on for. 'We're at the Omicron variant and I just feel this sense of dread – there are still a lot of letters in the Greek alphabet to go until we reach Omega...' she said, in between her tears.

It was shocking for me to hear this kind of unscientific drivel and on a mainstream radio channel no less. Does she honestly think that as clever a virus as Covid will stop varying itself once it reaches Omega? I felt like ringing in and putting her to rights. After all, Omega might be the last letter in the Greek alphabet but there are still many other alphabets to which the virus can move its attention. What about a script even more ancient than the Greek alphabet such as the Phoenician? Alep, Bet, Giml, Dalet, He, Waw, Zajin...... I can just imagine the BBC news report now: 'Welcome to the news at 9. The health minister announced today that the first case of the Zajin variant has been found in the UK. Margaret, a pensioner living in Barnstaple, found herself with a runny nose and immediately tested herself. "I was surprised, to be honest, that I have a case of the Zajin as I thought I would be immune ever since I was triple vaccinated for Giml last year."' Personally, I can't foresee a future in which Covid ever stops varying itself so, after the Phoenician alphabet, why not use Chinese characters, out of homage for the role which that great nation has played in its origins? Chinese has over five thousand characters so that will give us some breathing room for at least a few centuries before we need to work out the next script to use.

I have to say that I, for one, am glad we have moved on from naming the variants after their initial place of discovery. We went through a rather nasty phase, for example, of calling Delta the 'Indian variant'. This led to a massive rise in support for right-wing nationalist parties everywhere who made all kinds of outrageously racist claims about Covid …. 'coming over here, taking up our hospital beds…. what's wrong with the English common cold, that's what I say', as one former Councillor for UKIP is reported to have said. I was glad when that particular nonsense stopped. Of course, those UKIPPERS got a taste of their own medicine when a variant was identified in Kent, one of their strongholds. You might as well have called it 'the UKIP variant'. Indeed, I was delighted when Macron then blocked all incoming arrivals from the UK. After all, what business does Nigel

Farage and his ilk, especially if they are coughing and spluttering away, have with being in Europe anyway?

Now, as for Omicron, I am fed up to my back teeth with all these Covid-deniers pointing out that it is an amalgam (is that the right word, Ed?)[1] for 'moronic'. This is extremely offensive to experts such as myself. There is absolutely nothing 'moronic' about the world's response to this pandemic: it has stemmed from the highest levels of scientific and rational thinking.

In any event, the anti-vaxxers love to point out that Omicron is less virulent and that this makes sense because, as a virus varies itself it tends to get less and less virulent over time, and so why on earth are we taking so seriously something at this stage no different from the common cold blah blah blah.

Look, let's just get this 'a virus gets less virulent as time goes on crap' out of the way once and for all. Now, even if this were true, we should all still take Covid very seriously. For example, say that by the time we are towards the end of the Phoenician alphabet, the virus is so benign as to be not only always utterly asymptomatic but even so as to confer health benefits, well, the fact is that it is still Covid and you never know what it might get up to next. Sure, one moment you could catch a benign strain, have never felt better, get loads of work done, etc, and the next moment a new variant comes along with the fatality rate of Ebola and you and everyone you know drops dead. I'm not saying that this will definitely happen but Covid is a sneaky beast and so you never know what it will conjure up next.... it's always best to be cautious.

Indeed, the other day didn't I read of something which sounded even scarier than a variant? 'Flurona', they called it, a crossover combination of Covid and standard influenza. Utterly terrifying. It's only a matter of time now before there is a Poliorona or even a Leperona outbreak, and then we will all be truly and royally screwed.

Anyway, I hope I have illustrated just how serious and how very scary the emergence of all these variants is. In fact, when these Covid-

---

[1] I think so, Oisín - Ed.

deniers mockingly call them 'scariants', the irony is that they are, at last, speaking the truth.

Having considered variants and the kind of lies that the nutcases will spread about them, let's now turn to a whole range of other, typical myths about Covid. Indeed, you don't have to look long in the loony-sphere to see claims that the virus is seasonal or that natural immunity can fend off the disease or even that it matters how much viral load someone is exposed to. When I was thinking of how to combat these pernicious ideas in this book, I remembered that, at the mid-point of the pandemic, I did a Zoom Q&A with some of the local residents of Termonfeckin in which all of these mistaken notions unfortunately reared their ugly heads. So I have simply pasted a transcript of that session here as, in of itself, it can successfully dispel such misguided perspectives....

A Q&A with Oisín

"Oisín: Evening everyone...great to see you all. So how about we just get straight into it and whoever wants to ask the first question, just fire ahead....

Q: Hello Prof. MacAmadáin, Miriam O'M. here. I hope you are well. Em... I was just wondering what information we have now on the latest infection-fatality rate. I was listening to Prof. Ioannidis and....

Oisín: Professor who? Never heard of him! I hope you are getting your information from reputable sources, Miriam?

Miriam: Well, he seemed reputable enough, Prof. MacAmadáin, but, of course, you'd be best placed to judge his ideas. Anyway, he's just published a paper 'The infection fatality rate of Covid-19 inferred from seroprevalence data' and....

Oisín: Oh Lord, there's always one, isn't there? And trust that it would have to be the very first question!

Miriam: I'm sorry Prof. MacAmadáin?

Oisín: Just get on with it!

Miriam: Ok, well, he suggests that the median infection fatality rate is around 0.27%...what do you think?

Oisín: Absolute nonsense! At the Termonfeckin Institute of Expertise, we've calculated the IFR to be closer to 34%. Stick to official sources only, Miriam! Hmmmm, not the best start. Next question?

Q. Sandra B. here, Prof. MacAmadáin, the local florist. Obviously my shop is closed but for when we reopen, I was wondering whether you think the Covid can hide in flowers?

Oisín: Oh great, that's more like the kind of question we're looking for tonight, folks.... Yes, it can hide absolutely anywhere, Sandra, so I'd recommend that, to avoid your petunias doing the work of the grim reaper, you stay closed until everyone in Termonfeckin is at least quadruple vaccinated.

Q. Hi Prof. MacAmadáin, Maureen R. here, we met at the Termonfeckin Tractor Race a few years ago....

Oisín: Oh sure, I remember it well, Maureen. Didn't your husband get a medal that day?

Maureen: Well, he didn't have the best race actually Prof. MacAmadáin, sure, his tractor never even managed to start up. But anyway, Prof. MacAmadáin, I wanted to ask you....um, now I was delighted to get all my vaccinations, of course, but I have to say that the day after my 3rd shot I had, well, a little heart attack I suppose you could call it, just an absolutely tiny one mind, nothing to write home about really. I suppose it is quite unlikely but do you think it might have, perhaps, been the v.....

Maureen R. has been disconnected.

Oisín: Oh dear, we seem to have lost Maureen, though probably just as well as I'm really not sure where she was going with that question. Next?

Q: Hi Prof. MacAmadáin, Joan here, well, unlike the last speaker, I'm no anti-vaxxer I'm glad to say and I just wish to point out how wonderful I think these vaccines are. Now, I've had five shots altogether so when I got Covid recently, I knew that I would be just fine. Indeed, I stayed calm when I went hypoxic and my husband had to call the ambulance. I then remained optimistic all throughout my ambulance ride even when I started to exhibit the early signs of respiratory distress. And sure when I arrived at the hospital and was immediately put on a ventilator, didn't all the nurses comment on my cheerful disposition. And, here I am, Prof. MacAmadáin and all of you, alive to tell the tale and that I am still here at all is thanks to the vaccines. Truly, I tell myself every day: 'Joan, it could have been so much worse'.

Oisín: Thank you for sharing that very moving story, Joan. Similarly, I have a good friend in Dublin whose fully vaccinated father died from Covid. He also told me how much worse he knew it could have been. Next question?

Q. Dia dhuit, Prof. MacAmadáin. This is Patricia. I wanted to ask about the summer. Am I right in thinking that the virus is seasonal? Can we be less concerned about it in the summer and maybe leave the house once or twice even if only into the back garden?

Oisín: Since our government brought in mandatory masks in the middle of summer, how can you even be asking this question?! Do you think they would have done this if they didn't think that Covid was as deadly in the summer as at any other time???

Patricia: Oh, I'm sorry, Prof. MacAmadáin, how stupid of me...

Oisín: Damn right. As for your garden, you can go there but wear your mask. Next question?

Q. Good evening, Prof. MacAmadáin. My name is Sheila L. and I am a local naturopath. I was wondering about what we could do to improve our natural immunity? For example, Vitamin D supplementation?

Oisín: Vitamin D supplementation?! Oh, of course, why don't you just go and do a Donald Trump and tell us all to drink bleach while you are about it?! Honestly, I'm really starting to wonder whether I should have done this Q&A at all given the kind of questions I'm getting. No doubt you also think your crystals will heal Covid... next!

Q. Hi, Prof. MacAmadáin, Sinead H. here. Thank you so much for sharing your expertise with us all tonight. I, for one, have always been thrilled to have your Institute in our lovely town and hope to send my own son to college there. Anyway, Prof. MacAmadáin, what I was wondering was.... if we are asymptomatic how likely is it that we might infect someone else without knowing it?

Oisín: I'm very surprised that you are merely concerned about whether you can 'infect' someone or not. The more relevant way of putting your question would clearly be: 'Can I kill someone with Covid even if I don't know I'm murdering them?' and the answer to that is a resounding 'Yes'!

Ok, I'm tired of ye all at this stage...time for one more question...

Q: Prof. MacAmadáin, Deirdre, the local post office lady here, so I've made the decision not to be vaccinated and....

Oisín: What?!!! And you handle all our post?! I'm calling the Guards at once! Where's my phone.... Garda Morrison, is that you? I need to inform you of a terrible crime!

Deirdre: I'm out of here! It was nice knowing you all!

Oisín: Quick, everyone, get to the outskirts of Termonfeckin, close off all the roads! Prevent her from escaping from the town! Form a posse!

Sandra: But Prof. MacAmadáin?

Oisín: Oh, what now, Sandra? Time is of the essence....

Sandra: Will this need to be a socially distanced posse, do you think?

Oisín: Um, yes, yes, I suppose it will. Ok, so everyone: please keep two metres apart when you are in your posse and.....

Miriam: But I don't think gatherings between different households are currently permitted, Prof. MacAmadáin?

Oisín: Oh, what?! Oh, right, right. Ok, everyone, form a socially-distanced posse but only from within your own household.

Miriam: In my case, that means I can only bring my 3-year-old son with me as there is no one else here? Is that all right Prof. MacAmadáin?

Oisín: Um…

Sandra: Prof. MacAmadáin….what if she escapes more than 5KM from Termonfeckin? Would pursuing her further count as an 'essential reason for travel' do you think?

Oisín: Truthfully, I'm not sure…

Patricia: And when it actually comes to capturing her, Prof. MacAmadáin, how can we do that if we have to keep two metres away from her at all times so as to keep us all safe?

Oisín: I don't know to be honest. Does anyone have a stun gun by any chance?

Maureen: Oh, Prof. MacAmadáin, I've just seen Deirdre flying past in her motor. I think she's escaped us.

Oisín: Look, ok, everyone. She's got away. We'll let the Guards deal with it. But let this be a lesson for us all, as to the kinds of people who may be living in our midst and without us being any the wiser. I think I'll end the meeting now, I hope you found my expertise of help and, for sure, there is much of a sobering nature to reflect upon."

Well, I hope that's shown you the kind of arguments you can make when people come out with certain truly thick statements about Covid. I'm only ashamed to admit that some of the people in my very own home town were of these opinions. The police never did catch Deirdre either…. she escaped off to Mexico where they let in the unvaccinated without so much as even the shortest jail sentence (I'll be dealing with Mexico and its Covidian shenanigans in more detail later in this book).

On the plus side, I am happy to say that one of the wackier ideas put out there about Covid was not brought up on that particular night but, given how utterly prominent it is, well, I've no doubt but that you've heard mention of it....and so let's now move our attention on to....

The Origins of Covid

Now this idea is of course utterly mad! Some people really do think that the virus was leaked from a lab specializing in research that modified viruses (so called 'gain of function' research), and that this lab in Wuhan had received funding from the US Government's National Institute of Health. So I was especially adamant that I needed to bust this particular myth once and for all. But when I spoke to one of my mates high up in the US government, he told me that I really shouldn't. To be honest, my friend, who asked to remain nameless, got a bit weird about it all, saying it was for my own good and that Guantanamo Bay wasn't the kind of place you wanted to end up. 'No', I said. 'The lab is in Wuhan. That's where I'd be going to investigate not Guantanamo.' 'I don't think that you received my meaning, Oisín. Look, just don't cover it in your book, ok?'

My friend was so insistent that I backed down. After all, I'll always respect a good chum's request. All I can say is that he must regard it as such a doolally idea that it's beneath anyone to even try to counteract it. So I'll just leave it at this, then.

Anyway, there you have it folks, we have truly set the scene on Covid, its dangerousness, the tricks it can come up with, the multifarious ways in which people suggest it just isn't that big a deal and so on. And, of course, we examined at the start of this chapter exactly why lockdowns are the best way to go. However, some people would be of the opinion that lockdowns negatively affect our societies in some way or other...total rubbish, of course, and that's why, in the next chapter, we will reflect upon how life under Covid not only hasn't been that bad but, if anything, has been really rather a ball!

CHAPTER TWO: THE MANY JOYS & BLESSINGS OF LOCKDOWN

Now, if you were to believe the anti-vaxxers, life has become awfully oppressive in these last few years and society is falling apart under the strain of 'draconian' lockdown measures. Well, that hasn't been my experience AT ALL! And so, in this chapter, I'll be tackling the myth that some kind of dystopia has been created for our children, our relationships with each other or even the outlandish idea that everyone wearing masks somehow renders us less human....so read on, MacDuff, as I believe the saying goes....

Masks, Glorious Masks

One of my daily joys has been going down to my corner shop and seeing the sea of semi-blue faces all around me, a visible sign of our joint commitment to protect each other. And I don't know about you, but I also find that masks make people downright sexy and so I often catch myself gawking at the drop-dead gorgeousness of passers-by. In fact, there was a Cardiff University study about just this point which demonstrably proved that face masks make people more attractive and, as you know well by now, I will always follow the science.

On the other hand, when I pass someone on the street who isn't wearing a mask, or if their nostrils are showing even just a tiny bit, I used to give them a good hard frown (verbally admonishing them isn't the safest bet as it would be more likely to impart some viral load, not that the chancers wouldn't deserve it). However, I realised they couldn't see my expression and I was quite upset about this fact. But then, one day, I hit on a solution! Now I carry a load of angry face stickers and I just slap one onto my mask whenever I wish to indicate my intense displeasure at someone.

Most people have really got into the spirit of things. Some people, such as myself, will wear three masks and a protective visor. Others just two but it is still a good effort. Some keep their babies masked and others mask up their dogs when taking them for a walk. My cat never goes for a walk of course but I mask him up whenever we have visitors. And, whenever I go for a drive, I'll wear a mask. After all, viral droplets could get in through the air filter.

So imagine my shock when I heard talk of a study from Denmark which found that wearing masks produced 'no statistical difference' in infection numbers between the masked and the unmasked! As is always the case when I encounter likely conspiracy theories, I immediately went to investigate the source itself and what I found was a paper with an awfully long title[2] written by a certain Dr. Henning Bundgaard from the University of Copenhagen. Reading it left me so livid that I just had to write immediately to the British Medical Journal in order to set this author to rights:

'Dear Mr. BMJ Editor,

I am still in a state of near apoplectic shock having read the so-called scientific paper by Dr. Herring Bumgaard. Her paper examined the difference in infection rates between one group asked to wear masks at all times outside the house and another which was not. Over a two-month period a group of 6,000 people were divided up with around 1.8% of the masked group becoming infected compared with 2.1% of the unmasked group leading to the conclusion that there was 'no statistical difference' between the groups.

These kinds of results are all very well but the real question is what kind of ethics committee worth their salted fish would ever give approval for such a study? Instructing three thousand people to wander around in the middle of a pandemic *without* a mask? Whatever the findings of this study might have been, the fact remains that it could have

[2] 'Effectiveness of Adding a Mask Recommendation to Other Public Health Measures to Prevent SARS-CoV-2 Infection in Danish Mask Wearers: A Randomized Controlled Trial'

resulted in wholescale murder on an unprecedented scale. In fact, I'm not so sure that it didn't. I mean, how do we know that up to 100% of those infected in the unmasked group didn't end up ultimately dying due to greater viral exposure? Were they only asked about whether they were infected and not whether that infection had killed them? Or, if they themselves didn't all die, perhaps they ended up wiping out whole districts of Copenhagen due to transmitting their droplets at much higher levels of viral load. Did Dr. Bumgås think of either of these possibilities? Why were these questions not addressed in her study? These and so many other things remain unanswered. I call on the Danish government to launch a public inquiry forthwith.

Like many Irish people, I have more than a few drops of Viking blood in me. Therefore, it saddens me greatly to see how some of the lands of my forebears have handled this pandemic. It's bad enough that Sweden decided to murder all their grannies without the Danes now coming out with disinformation in the guise of a scientific study, however credentialed Dr. Mackerel might say she is.

Is mise,

Prof. Oisín MacAmadáin, Termonfeckin Institute of Expertise.'

And on the very day I sent off this riposte, didn't I read in the Irish papers that a group of right-wing parents were gathering to protest children wearing masks in primary school? It's studies like those of Dr. Bümflüff that end up fuelling these far-right extremists. But such researchers never seem to be troubled when they reach their conclusions, not only by the fact they don't accord with the science but also because they will encourage the anti-vaxxers.

While Dr. Bumfårt mightn't be following the science, I most certainly am. Only yesterday I read a study from Cambridge University (FAR more prestigious than Copenhagen University though obviously less so than The Termonfeckin Institute of Expertise, even if I say so myself), a study which I really admired for its efforts to work out how masks can help us save even more lives than they already have. It was called 'Face mask fit hacks' and what it found was that if someone wears a section of pantyhose over their mask, they reduce their viral load by a whopping SEVEN times. There is nothing like real scientific research to work me up into a state of excitement and so I immediately put on my mask and rushed upstairs to my wife. 'Get your pantyhose off now, dearest!' 'Oooh, Oisín', my wife responded. 'I so love it when you surprise me like this…. here you go, oh, but what are you doing, Oisín?'

As I stood admiring my new look in the mirror, I knew that this scientific breakthrough by the clever clogs at Cambridge NEEDED to be adopted nationwide as a matter of urgency. And so I knew *exactly* what I was going to talk about next time I popped onto RTÉ, our national broadcaster. After all, have they not been trailblazers when it comes to discussing innovative solutions to this pandemic? I remember well seeing a couple of fellows on one of their shows, and one of the chaps no less than a professor of biochemistry, standing inside large protective bubbles, saying, quite rightly in my view, that they had found a way for people to go to concerts safely. So I know the editors on RTÉ will be receptive to this idea…. between the giant bubbles and the pantyhose, we'll get there.

A Short Word on Dating

All this talk of stockings has put me in a suddenly more amorous frame of mind and even though I am, of course, happily married to the missus, if I were a young fella playing the field, this is how my Tinder profile would read (I hope it will serve as a helpful template to any youngster out there who might erroneously think that the lockdown has ruined their chances):

"Username: Oisínsexyness

Age: 22

Looking for: Female, 18-30, *at least* triple boosted.

Vaccination status: Double vaxxed and three boosters.

Hey darling! I've been told that I'm destined for great things. Let's organise a Zoom and swap notes on our vaccination experiences! As long as you don't believe Bill is out to chip you, LOL, then I'm sure we'll get on just fine.

Finally, if we do meet when lockdown is over, I want you to know that I always wear protection and I hope that you'll also always wear your mask."

Honestly, with profiles such as this one, you'd be unlikely to go too far wrong.

But what about the supposed adverse effects of lockdown on our children? Has their education or mental health suffered as a result of being asked to play their part in the war on the virus? I very much think not! And so....

No, the Pandemic is NOT Leading to Mental Health Problems in Children

I can't for the life of me understand all these anti-vax, far-right parents and their 'concerns' about the mental health of children during the pandemic. They get all worked up about what effect lockdown restrictions, school closures, mandatory masking in schools and so on, will have on their precious little angels. This is mollycoddling, pure and simple. No good ever comes from wrapping your children in cotton wool and pretending that reality doesn't exist. And, anyway, these children are far more resourceful than their mentally ill parents realise.

One really does wonder what the problem is. I mean, imagine that you are your six-year-old self and being told that....

- a deadly virus is spreading rapidly throughout every country in the world

- you could be carrying this deadly disease at any moment, even though you'd have no symptoms and therefore wouldn't know it, and that you could therefore kill your grandparents and so you can no longer see them ever again, or at least not until the wonderful day when they are vaccinated

- and not only that you might inadvertently kill your grandparents, but that you might even kill your parents so it's best at the very least to drop hugs and kisses and keep well away from them too until the wonderful day when they are also vaccinated

- but that it is not something that is likely to kill you, thankfully (rather, you are just very likely to be responsible for the deaths of practically everyone else, at least until the wonderful day when they are all vaccinated)

- and that you should wear a mask in school at all times so that you play your part in making sure you don't kill not only your own parents and grandparents but also all of your classmates' parents and grandparents as well

- but that you then learn it seems there is a slight but real chance that the virus might kill you after all and so you need to wear the mask to protect both yourself and everyone else in your class, otherwise you might kill them or they might kill you, at least until the wonderful day when you all get vaccinated

- and then you learn that, even though the wonderful day when everyone gets vaccinated has arrived, that all of the above still applies

Now, I ask you, how on earth could this kind of scenario lead to children experiencing mental health problems of any kind whatsoever?! To my mind, the only outcome from this kind of clear messaging is the development of responsible, resilient and empathetic children. Each day, as they set off to school, they are well aware that they could be infected by an invisible enemy and potentially kill anyone they pass on their route. Does this not naturally encourage them to respect people's boundaries? And when they are in school, they are similarly careful to avoid contact or communication of any kind with all their classmates, knowing that to do so would run the risk of never seeing their friends again? Is this not, as I believe they say, the very acne of empathic care? And, finally, they are aware of the huge personal risks they are taking every day, in that they too could be killed by the virus at any time. Truly, they are our little heroes.

Indeed, there have been many unexpected blessings of the pandemic and one of them is that we can look forward to the next generation being altruistic and calm in the face of adversity. Just, of course, like we have been throughout this pandemic. For is it not the case that children ape their elders?

So let us pat ourselves on the back too. The adults in the room (among whose number I do NOT count the infantile anti-vax parents that I have described above) are truly leading by example.

The Ideal Covid Classroom: A Case Study

Now, as for the idea that children have received something of a subpar learning experience over the course of the pandemic, the fact is that there have been plenty of schools which have risen to the challenge and nevertheless provided as good, if not better, an education than before the Covid era began. And so when I was thinking about how to rebut this particular nonsense, my dear old friend, Ms. Gretel Voopingkoff, came into my mind. Gretel is a primary school teacher in Germany, a place which, to my mind, has excelled in its handling of the pandemic. So I got in touch with her by email and asked her to send me an account of what school life there has been like in the time of Covid. What I received back should convince anyone that Covid really can lead to a classroom atmosphere in which the education of our young folk can truly flourish. Here is her email:

"Hey Oisín!

So wunderbar to hear from you again! I have been following your awesome work in exterminating anti-vaxxer propaganda! Let's hope it is not too long before we will not have to put up with them anymore.

As to your question about school life here, it is also so wunderbar......

The day starts at 8AM and each child comes to the front of the class and states their current vaccination status, whether it is 'ein jab', 'zwei jab' or 'drei jab' (so called 'super duper triple booster'). Those with one jab are given polite applause while encouraged to amend their ways, those with two jabs are clapped enthusiastically and those with three jabs.... well, we all play the geese marching game (an old game we have here in Germany, a bit like you with your ring around the rosy) and salute each other in the traditional manner. However, if they say they are unvaccinated, we all stay deathly silent und stare at them.

Sadly, there are still four unvaccinated students in my class (their parents are 'antivaxxers', ugh!) and so we have partitioned the room and make sure to keep them separate at all times. Furthermore, while all vaccinated students need only wear one mask, the antivaxxer children must wear three masks as well as Hazmat suits. This means they can't hear me and therefore learn anything but really the main thing they need to learn is the error of their ways.

We also have a new alarm system in place which is able to detect very specific noises, namely sneezes und sniffles. If a student sniffles, the alarm system goes off throughout the whole school: 'Achtung, Achtung! Suspected Viral Presence!' Then all students must stay exactly where they are while our specialist Covid Protection Unit identifies the location of the sniffle. Then, that student, and everyone else in their class, whether vaccinated or not, is taken away to a special camp.

So, all in all, Oisín, teaching is the same delight it ever was!

Yours affectionately and with an electronic (and therefore socially distanced) hug,

Gretel"

This is truly wonderful stuff which just shows us what can be done, does it not? Indeed, how could anyone feel that such a classroom atmosphere could provide anything other than an absolutely brilliant learning experience? In order to be able to learn effectively, our children must feel safe and able for the challenges of this new kind of world, and this sort of approach ticks both of those boxes. Indeed, we all need to copy this model! And then all but the most recalcitrant antivax parents would be forced to admit that the general education system has gone up more than a few notches and that if they wish for their children to benefit and be out of those Hazmat suits…. well, it's jabby jabby time. But to be honest, I doubt that they'd ever even recognise such improvements…after all they are the kind to be in the gutter while the rest of us are all starry-eyed or whatever it was that George Bernard Shaw once said.

So there you have it, folks….no sane person would ever suggest that our daily lives, or the lives of our young offspring, have been affected adversely by the pandemic. Masks show us how much we all care for each other and our children are growing up in an atmosphere of societal love and compassion unlike anything ever witnessed before. So, let's all cheer for Lockdown and the gifts it has brought us! Hip hip, hooray!

But how, dear reader, can you deal with antivaxxers when you encounter them in your everyday lives? For example, when you are going into a coffee shop only to pass someone who murmurs 'Lucky for you to be allowed in there!'…. what kind of comeback should you make? How, indeed, should you 'factcheck' them? Well, this is the focus of the next chapter…and so on we go!

CHAPTER THREE: OISÍN'S GUIDES TO…..

…Factchecking

The great problem with the internet is that any old nutbag can write and publish whatever the hell they want (and, indeed, that's one of the reasons why I was motivated to put together this book).

And, when it comes to the pandemic, misinformation is out in full force, spreading ever more quickly and dangerously than Covid itself. Luckily, the good sports at all the major tech companies were wise to the danger early on and trained up whole armies of Misinformation Spotters & Factcheckers. These sterling chaps and chapesses put the record straight on all the myths that are getting put out there. And, boy, are they smart. I don't know quite what kind of training they have but it wouldn't surprise me if doesn't require, at a bare minimum, gaining a PhD in virology or some such. Anyway, the point is: these guys really know their stuff, of that we can be sure.

And we should all be grateful for their efforts. Did you know that, as of going to press, YouTube has taken down over 1 million videos spreading Covid misinformation? That's 1 million videos extolling the virtues of bleach or the 'dangers' of 5G that will now never subvert those who have the more suggestible and softer minds among us. Or that Facebook has taken down countless whining antivaxxers who claim to have had a heart attack or died or whatever after their vaccination? Like the group of 120,000 such right-wing extremists which was deleted just like that.

But here's the thing: you don't need to be a top expert like myself or an employee of Facebook to know how to factcheck things. Indeed, anyone can do it. Let me give you a quick guide so that you can too can factcheck the Covid-deniers in your life.

In general, factchecking involves doing one of the following three things:

1. Pointing out that the EXPERTS disagree with the misinformation
2. Pointing out that the person spreading the misinformation is, in fact, a crackpot
3. Pointing out that, even if the misinformation is correct, it still isn't true

Let's now have a look at the following examples. See if you can spot which strategy I am using in each factcheck (sometimes I use more than one!). I hope these can serve as templates which you can adapt to any situation you see fit.

" Anti-Vax Claim No. 1: Lockdowns cause more harm than good

Factcheck: A typical claim by opponents of lockdowns is that they cause more damage than they prevent, in particular by causing harm to the economy, livelihoods, mental health and the provision of care for other health conditions. However, EXPERTS at the University of EXPERTISE, located in EXPERT LAND, take a different view. For example, their study, *Lockdowns Predict Increased Positive Mental Health Due to Augmentation of Time Spent Loafing and Eating Takeaways: A Qualitative Analysis*, indicates that lockdowns have actually led to increased happiness levels across the developed world. Meanwhile, Mr. Extremely Smart at The Intelligent University suggests that.... and so on.

Anti-Vax Claim No. 2: The inventor of the mRNA vaccine technology, Robert Malone, says there are safety concerns with the mRNA vaccines

(Note also the use of footnotes and quotation marks with this strategy – these can be very helpful)

Factcheck: Dr. Robert Malone, a former scientist and now full-time anti-vaxxer, 'claims' to have 'invented' the mRNA vaccine technology while at grad school, although this is disputed.[3] He is known for spreading vaccine misinformation to the point that Twitter removed his profile. A research study from the University of WMS (Woke Medical School) concluded that having a beard, white skin, and maleness, all characteristics that Dr. Malone happens to share, are strongly associated with the risk of becoming an anti-vaxxer. In contrast to Dr. Malone's view, most experts state that the mRNA vaccines are extremely safe and should be taken by everyone from birth onwards at least ten times.

Anti-Vax Claim No. 3: VAERS (The Vaccine Adverse Event Reporting System) shows that over 29,000 people have died following Covid vaccination as of the end of June 2022

Factcheck: The use of VAERS data to support the idea that the Covid vaccines are dangerous is a common tactic used by anti-vaxxers. VAERS is a self-reporting system, however, and not subject to the rigorous scientific measures employed in clinical trials which have shown the vaccines to be safe and effective. The reports of over 29,000 deaths in the VAERS database do not prove that the Covid vaccines are dangerous: they merely show that 29,000 people happened to die shortly after their vaccination. Death is a statistically common phenomenon which experts have found to occur in most populations. Therefore, it is unsurprising that, in the context of a mass vaccination program, a small number of people happen to die in the days following their vaccination.

Anti-Vax Claim No. 4: Certain groups are more at risk from Covid than others

Factcheck: Covid-deniers typically claim that the aged, infirm and those with certain underlying conditions such as diabetes and heart disease, are most likely to have a negative outcome from Covid. While the median age of death from Covid is 83 years old and less US children and adolescents have died from Covid than from standard

[3] LINK NOT WORKING.

influenza, experts nevertheless disagree. Prof. Nadir Jibjab states: 'Let me be very clear.... there is no group that is not 'at risk' from Covid: the old, the young, infants, even foetuses can all have a very, very, very, very, very BAD outcome'. One of the principal problems of anti-vaxxers spreading the above idea is that it promotes the concept of 'risk stratification', or the idea of identifying the particular health risks that different societal groups face from Covid-19 and developing public health strategies which are appropriate to each group. Not only is this approach discriminatory and ageist, it might also lead some people to doubt the value of locking down everyone. Prof. Jibjab continues: 'It is very important that people understand that shutting down all businesses, cafes, restaurants, not seeing anyone else or ever leaving your own home, are all tools based on the strongest science.'

Anti-Vax Claim No. 5: Improving natural immunity is an effective way of reducing a negative Covid outcome

Factcheck: Many anti-vaxxers claim that a healthy immune system is capable of fighting off Covid. This kind of 'natural health bias' likely stems from the fact that anti-vaxxers are usually interested in alternative and poorly regulated health cures such as earthing, crystal healing and colonic cleanses. Experts, however, point to the fact that the immune system, outside of the context of being vaccinated, is no longer as central to the body's health as was once believed. Indeed, many who have died from Covid also had immune systems. Most experts believe, therefore, that it is best that everyone just gets vaccinated at least once every three months.

Anti-Vax Claim No. 6: Not all Covid deaths are caused by Covid

Factcheck: A common idea spread by anti-vaxxers is that Covid death numbers are exaggerated due to the way in which they are counted. For example, someone who dies purely of cancer may get recorded as a Covid death even if they only had a mild or asymptomatic Covid infection in the previous month. In these kinds of cases, anti-vaxxers claim, deaths 'with Covid' should be distinguished from those 'of Covid'. Prof. Hubert Müzzleup from the Institute of Zero (as in *Absolutely* Zero) Covid Studies, however, takes a different view: 'This is absolutely outlandish. Take even a scenario where someone dies from falling off their bike…. how do we know a Covid-induced sneezing fit didn't lead them to lose control of the handlebars and therefore make a directly causal contribution to their demise? Covid is capable of all sorts of tomfoolery and, on this point, the science is quite clear, let me tell you."

So, there you have it, folks, Oisín's guide to factchecking. It truly is a breeze. So, next time you are faced with the kind of ludicrous things the Covid-deniers come up with, simply adapt one of these strategies and you are good to go. They'll be left speechless, that I can promise you!

But, of course, I recognise that not all of my readers will be your average Joe (or your average Brandon for that matter, not that *he* is in any way average!). I have no doubt but that some of you must be journalists at the most prestigious newspapers, such as *The New York Times* or *The Washington Post* and, of course, I know for sure that all my staff at *The Termonfeckin Tribune* (of which I am the editor) will be reading this book so as to make sure they keep their standards up. Therefore, I also wanted to include a guide for those amongst my readers who are shaping opinion at a societal level…this next section is for you!

And so on to my guide to….

…Covering Covid in the Media

Ok, let's cut to the chase. There follows a list of seven golden rules for covering the pandemic (of course this list isn't exhaustive, in truth there are hundreds of rules, but lest this book become a mighty tome, I narrowed them down to the most essential).

Rule 1: When a government lifts, or so much as suggests lifting any restriction(s), point out that experts everywhere think this is a *terrible* idea

Example:

'Boris Johnson has announced plans for UK's 'Freedom Day', when all Covid restrictions will be lifted, and the UK will move into a new phase of 'living with the virus'. Experts, however, are (insert one of the following depending on desired effect) *advising / warning / pleading / begging on their hands and knees* for the Prime Minister not to take this course of action. A letter has been signed by *241 / 4,300 / 2.1 million experts* who have predicted that 'Freedom Day' will result in *500,000 / 18 million / everyone in the UK and whole world* becoming infected within weeks along with a *significant / enormous / truly biblical* death toll. Dr. Smärtz Aleks, one of the letters' co-signatories, said "What does he mean 'living with the virus'? The very notion is absurd. You don't live with this virus, you just die from it, and that's that."

Rule 2: When a government lifts, or so much as suggests lifting any restriction(s), point out that ordinary people everywhere *also* think this is a *terrible* idea

Example:

'Boris Johnson has announced plans for UK's 'Freedom Day', when all Covid restrictions will be lifted, and the UK will move into a new phase of 'living with the virus'. Reaction on the streets of Exmouth in Devon, however, was one of concern. Miriam, a retired hairdresser, was less than happy with the Prime Minister's announcement. 'Well they might as well just send in the army to murder us all and get it over with!', she said. This view was echoed by Tim, a local Councillor: 'But we haven't had our sixth booster yet. How can this decision possibly be safe? It's madness!'

Rule 3: When anti-vaxxers protest in any way shape or form, make sure to highlight just how fringe, conspiratorial, odd, potentially dangerous, small in number, utterly selfish and generally unrepresentative of the mainstream they are. In these kinds of cases, it is also helpful to make liberal use of quotation marks.

Example:

'The invasion of Ottawa by truckers campaigning for 'freedom' has reached its 11th day. A total of 12 truckers were present. One of them, Margery, who had purple hair, a cat under each arm, and was waving a placard with the words '5G is the vaccine!', said 'We are just peace-loving Canadians who want our country back!', before proceeding to take out a revolver and fire some shots in the direction of government buildings. The Canadian government is facing increasing pressure to act decisively over the 'freedom' convoy, which says it also campaigns for 'bodily autonomy' and 'inalienable human rights'. The protest comes at a time when gatherings of more than 1 are still illegal and, therefore, may constitute a super spreader event that will ultimately threaten the progress the Canadian government has made in trying to save lives.'

Rule 4: Never, ever publish a letter from an anti-vaxxer in your newspaper. As soon as you know it is from one, don't even read the rest of it…. instead, put it straight away into a separate folder entitled 'cranks.'

For example, *never* publish a letter that starts something like this:

'Sir,

We, at the Centre for Civil Liberties, are increasingly concerned by the government's removal of basic rights as guaranteed by the Constitution….'

But *do* publish:

'Sir,

I was horrified to go to the hairdresser the other day and learn that my hair was being cut by someone who is unvaccinated (and, worse, proud of it!). I kept my mouth shut so as to avoid inhaling any of her droplets and am currently sitting in my car outside the nearest A&E so that I won't have far to go when the moment comes. My hairdresser should be ashamed of herself and I feel that she, and her kind, should be locked up forever rather than potentially murdering us all.

Yours, etc.

Maggie O'Muirahertaighach.

In her car outside St. Jimmy's Hospital, Dublin.'

Rule 5: When you absolutely *must* interview an anti-vaxxer on your radio or TV show, make sure they meet at least five of the following criteria:

1. They practise a holistic health profession, such as aromatherapy or Reiki, and believe that this cures cancer

2. They are a Christian

3. They express the belief that the US elections were stolen, that the Capitol Hill riots were organised by Antifa and that Trump is simply downright marvellous

4. They have an internet trail to fringe websites on which they have voiced support for Nazi ideas and theories. Ideally, they have a swastika tattooed in a visible location.

5. They believe that the world is run by an elite group, 'The Illuminati', who are probably aliens

Rule 6: Whenever describing someone who claims to have suffered a 'vaccine injury', make liberal use of quotation marks. Make sure to include a statement to the effect that they are nevertheless still thrilled to have had the jab (do whatever you have to do to make them say this).

Example:

'Candy, a 26-year-old woman from Dallas, 'claims' to have developed 'Guillain-Barré syndrome', a 'rare neurological disorder' after her second vaccine. She believes her symptoms include episodes of 'near total paralysis', 'severe muscle weakness' and 'difficulty swallowing'. She says that her initial reaction was 'life-threatening' and 'required hospitalisation for 3 weeks'. Nevertheless, she is still delighted that she got the vaccine. 'What are you doing?! No, of course, of course! I'm so delighted to have got the vaccine! Over the moon, in fact! I mean, the Guillain-Barré side of things, that's totally manageable really and, what with omicron coming at any moment, I'm so glad I'll be spared the worst of that…. can you please put that gun away now?'

Rule 7: When discussing incidents of increased excess non-Covid mortality among the young and middle-aged, make clear that these cannot be attributed to the vaccination program but to just about anything else.

Examples:

'Experts warn that increased rate of heart attacks among middle aged men is due to increased anxiety about having a heart attack among that group'

'Lack of proper oral hygiene likely cause of increased rates of heart inflammation among young men, says expert (and who are we to doubt him)'

'Leading researcher suggests rising numbers of strokes can be attributed to increasing pet allergies'

And so on.

***

Rightio, that was my guide to talking about Covid in the press. In a way, I don't know why I wrote it. After all, the media everywhere have been following rules of this kind all along. But what I do know is that we are in a battle of ideas and we all need to play our part, whether, at one end of the scale, that might mean factchecking someone on one of your WhatsApp groups or, at the other, writing an op-ed for *The Guardian*. Indeed, if we look throughout history, it is always the side that strives to stop the thickos in the room from speaking their point of view that ultimately wins. And we need to win now too for, Lord help us, people these days can truly be thicker than ever.

Look, I wouldn't blame you if you weren't a bit downhearted after reading this chapter. I mean, isn't it depressing that we even need to tell people how to think in the first place? Why is it that some people just aren't quite as mentally connected as the rest of us? So, it's time for some truly good news at this point as we move towards our next chapter. Indeed, thankfully most of the government cabinets around the world are very clued-in and have responded to this pandemic in the most scientific ways possible. So let's now treat ourselves to the crème de la crème of these last few years and The Lockdown Hall of Fame!

First up, the land of saints and scholars, my very own beautiful country, the Emerald Isle....

CHAPTER FOUR: THE LOCKDOWN HALL OF FAME

Ireland: A Model Case Study – Probably the Best Lockdown in the World!

Like all proud Irishmen, I believe Ireland is pretty much the best at everything and our approach to Covid-19, or 'the Covid' as we affectionately call it here, has been no exception.

At the mid-point of the pandemic, I remember well watching a video of the valiant members of An Garda Síochána (that's our national police force for those among my readers who haven't yet had the pleasure of learning our beautiful native tongue) arresting an Evangelical pastor in Dublin during his Sunday service. Now, this pastor was shamelessly breaking the Covid laws which prohibited all in-person religious services at that time and yet, in the video, didn't he go on and on about his constitutional right to worship or some such nonsense along those lines. The cheek of him and I'm glad to say that the Guards didn't put up with any of his carry on: into the van he went and off to the nearest jail. I mean, what was the man thinking, putting us ALL at risk? Sure only the day before, two people had died of the Covid.

Bizarrely, some people at the time criticised this criminalisation of communal worship, pointing out that it puts Ireland on too close a footing with the likes of Saudi Arabia or North Korea. Honestly, how racist can you be?! I, for one, think that the Irish government made the 100% right call in this instance. Sure we all know that these religious types can get a bit carried away, jumping around and singing praise at the top of their lungs…. Indeed, I'd say that it is only a matter of time before a study confirms that they are arguably the worst spreaders of the disease (after the unvaccinated, that is). Well, that's among the evangelicals anyway but even among the Catholics, who tend to be quite reserved in Ireland, it's clear that Mass is a public health disaster waiting to happen. Think of that part where they all toddle up to the priest to have their holy communion. What if the priest has Covid? Can you imagine the headline? 'Single priest kills entire Parish in Shock Super Spreader Event'. No, in my mind there's no doubt but that online worship was the only way to go.

In any event, the response to Covid by Irish officialdom filled me with pride from the get-go. Indeed, right back in March 2020, the government created an advisory group to work out the best way to respond to the pandemic. This group is called The National Public Health Emergency Team, or NPHET (but everyone pronounces it like NEPHET which sounds a bit ancient Egyptiany and therefore rather cool, you know, a bit like Nefertiti or Nebuchadnezzar. Now, having said that, it isn't the acronym I'd have chosen. You see, I'm rather a fan of acronyms which also mean something on their own, like the SAGE group in the UK [the 'Scientific Advisory Group for Emergencies' and a very wise bunch they are too]. In hindsight, perhaps something like COMPLY would have been better ['Committee for Overseeing this Monstrous Pandemic and the Lies which Yobbos will come up with about it'] but we'll just have to bear that in mind for next time.)

In any event, NPHET have done a great old job. Indeed, every Covid regulation they came up with was based on the greatest scientific thinking. For example, citizens could only travel up to 5kms from their homes for most of the pandemic, unless for an essential purpose. Now, if you weren't the brightest spark, you might wonder whether someone driving, say, 9kms from their own home would pose a public health risk.... there they are in their car, a confined and closed off space and so on and couldn't they follow social distancing guidelines wherever they might get out, blah blah blah. But what if they were to open their window at some point and, being suddenly overjoyed at the novelty of their surroundings, begin to belt out U2's 'It's a Beautiful Day' from the top of their lungs only for a Covid-infused droplet to carry on the wind and land on a little old lady standing at a bus stop who, very shortly afterwards, is then carted off to ICU...do the doubters ever even entertain such a scenario? No, of course not and that's why we should all be grateful to have had NPHET making it damn clear just what it was that we all had to be doing.

And this distance rule applied equally to all, even to those living in the countryside. The very gumption of some country folk to think that they should have had a special exemption from the 5KM rule! Oh yes, sure you live in an isolated spot and there's not another soul in sight and therefore you think the risk is minimal and sure why can't you go to the beach which is 8KM away…. you selfish bastard. Well, we are all aware that country folk tend to be a bit simple. For all we know, the science will probably show that Covid can be carried by sheep, horses or other animals that this lot tend to be around all the time and that next up there will be Crazy Sheep Covid Syndrome (CSCS), which will be even worse than Mad Cows Disease. Well, we don't know for sure and so we have to plan for every eventuality and, in that light, the distance rule clearly makes as much sense in the country as it does in the city.

But while I've had the cockles of my heart warmed by the government, this is nothing in comparison to the admiration I feel for the Irish media, in particular RTÉ, the national broadcaster. They have done a sterling job of reminding all of us about exactly what kind of divilment Covid was up to at any given moment. So much so, in fact, that I remember one morning when, totally innocently, I said to the missus 'Darling, will you turn on Covid Radio 1 for me please?' 'Covid Radio 1, Oisín? Do you mean RTÉ Radio 1, dear?' 'Oh yes, how silly of me!'

Anyway, the standard of their news coverage has been simply phenomenal. From memory, here is an example of one of their reports:

> "Welcome to the news at 1 with Sharon Ní Baol-dom. NPHET have reported a worrying rise in the number of Covid cases among teenagers and children with 55% of the 2,641 latest confirmed cases belonging to that group. We now go to our reporter, Cormac Scaoillmhóir. Cormac?
>
> Yes, Sharon. I'm here now at Our Holy Lady of Lost Causes school in Drimnagh and what an apt name that is too after this latest tragedy. Here, 3 of the 123 pupils recently tested positive for the disease and so the school board has decided

to close the school for at least the next 3 months. I'm joined here by Séamus, a friend of someone who contracted Covid-19. Séamus has had a negative PCR test prior to this interview. Séamus, tell us about your friend, Rory:

Yeah, well, Rory, he got a bit of a sniffle, like.

And was it a bad sniffle, Séamus?

Yeah, he was full of gunk coming out of his nose. It was minging.

I see, that sounds very bad. Well, Sharon, as you can hear from Séamus, this is not just a disease which can affect the very old, but also the very young and otherwise healthy. Concerns around the safety of children are rising amongst parents who are increasingly calling for the closure of all schools for the foreseeable future. Back to you in the studio."

This is ruthlessly brave reporting, plain and simple.

And, to return to the powers that be, our government also made some ruthlessly correct decisions, hard judgement calls that turned out to be utterly spot on. For example, at the mid-way point of the pandemic they introduced mandatory hotel quarantine, with arrivals into the State from specific countries forced to undergo a two-week period, at their own expense, in a hotel just outside Dublin airport. Now, mind you, they didn't just introduce this rule for arrivals from any old country or places you might expect like France or Germany... no, they focussed their efforts instead on the places you wouldn't necessarily assume to be the greatest harbingers of the disease to Ireland's shores. But if our government agents felt that places such as Angola, Rwanda and Colombia, and other such countries which have no direct travel link to Ireland, were the most likely to send forth droves of sniffling hordes in our direction, then I have no doubt but that they were following the science. And furthermore the government, acting correctly on the principle that caution is king, was also wise to the possibility of world-threatening outbreaks from even the most unlikely sources - tiny countries such as Monaco, Andorra and San Marino. Sure, in San Marino, a country of 33,000, they were typically having all of 25 cases in a week. That's 25 people who might have otherwise decided that now was the time to visit Ireland. Such actions save lives. And also, we haven't forgotten how San Marino once scored a goal against us in the football, their only in the history of the sport, I believe. Revenge is sweet.

In any event, my conclusion regarding the Irish pandemic response is that it was truly marvellous altogether.

There were some blots on the landscape, though. For example, a rather nasty piece of work by the name of 'Ivor look at the data Cummins' who was spouting conspiratorial nonsense on YouTube every week. Oh what I wouldn't have given to 'debate' him with a sliothar! And then, of course, we had a whole range of far-right extremists protesting in Dublin. This was not only unsporting behaviour on their part but also put all the Guards in danger. I mean, the vaccine roll out had not even begun at that point and therefore the Guards couldn't club any of them without putting their own health at risk. But sure then, these kind of people can never think beyond the skin of their own noses.

Anyway, there are bad eggs everywhere but, in general, the Irish people have been very much clued-in, scientifically speaking, throughout this pandemic and I call upon all governments everywhere to copy our approach which was arguably the best in the world. Well, we sure do tend to punch above our weight in just about everything we do and it has been no different with the Covid.

Now, all that said I must admit that one part of me feels a little insecure about the Irish pandemic response. Just a teeny bit, that's all. You see, before long I also learnt that there was a whole other model for dealing with Covid, one that sounded as close to Utopia as one could imagine (a 'Covidopia' as it were) and I felt somewhat ashamed that we hadn't adopted this way of going about things in Ireland, as terrific as our efforts were, of course. You see, this other approach, leads to a world where no one is ever sick from Covid, no one ever dies from it and where everyone stays safe forever. They call this approach the 'Zero Covid' way and, in essence, all it involves is for everyone in a country (and ideally the whole world) to avoid contact with everyone else for the rest of their lives and get vaccinated every four months.....but simple and utterly failproof as this strategy is only a few brave places actually adopted it. Anywhere that did so has my eternal and undying admiration and one of those places happens to have been our next country in the Lockdown Hall of Fame, namely...

Australia (hmmm might just have done the whole lockdown thing a bit better than us actually)

I will never forget the day when I read a report in *The Oirish Times* about how a security guard in Perth had tested positive for Covid and then the whole city of two million people was immediately locked down. From that moment on, I was hooked. 'Now there's a country which takes the Covid seriously', I thought to myself and I immediately set about researching as much as I could find about Australia's approach. What I found was truly a land of milk & honey (or of 'vaccines and masks' I guess you could say). Truly, the Aussies excelled in every way when it comes to Covid and to do justice to their achievements would require a whole thesis. So I've decided instead to home in on their vaccination program, from which you will get a sense of their whole generally magnificent approach...

The Aussie Vaccination Program

Now, I often wondered how exactly the Wallabies would tackle the Vaccination Question.... would the unvaccinated be (rightly) banned from any communal didgeridoo playing? A few months ago, I watched a TV news report from The Land of Oz which revealed all:

> 'Hi Shane. This is Sheila and I'm out in the streets of Melbourne where the latest health laws just came into effect. The unvaccinated remain barred from entering most indoor venues but, from today, if they are caught coughing or sneezing in public, they can be arrested and imprisoned for up to six months. Citizens are encouraged to watch out for anyone showing signs of viral activity, even subtle signs such as a runny nose, and to report these to their local police. I'm

joined now by district chief of police, Mike Giblet. Mike, can you tell us more about these new regulations?

"Yes, Sheila. From the first day of this pandemic, we have all become accustomed to watching out for typical signs of viral activity in ourselves and our immediate loved ones. But now we are being asked to extend that same vigilance outwards towards others: if you spot anyone coughing or sneezing, particularly if you have strong reasons to suspect they are unvaccinated such as an odd appearance and demeanour or the wearing of a tinfoil hat, please ring our dedicated helpline RUD (Reporting Unvaccinated Dangers) on 155 and file a report. We will take the matter from there."

Thank you very much, Mr. Giblet. In the interests of balance, we are now joined by *spits* an unvaccinated person, Michelle. Michelle is miked up and standing at the other end of the square so as to keep us all safe. Michelle, are you not a disgusting person?

"No, I am not.... I am simply exercising my bodily autonomy and have decided that, given I have a history of anaphylactic reactions to vaccines..."

Anaphylactic reactions? Nothing that an EpiPen can't deal with, surely? How can you be so selfish?

"Well, these reactions can be fatal and..."

Well, so can Covid, so there!

"Yes, but *cough*, sorry I have something in my throat..."

What was that? Did you cough?!!

"Yes, I had something in my throat, but anyway, as I was saying.... hey, what's going on?! Leave me alone!!"

And we are now watching live as one of the first arrests under this new legislation is being carried out by Melbourne's finest. We can see that Michelle is now surrounded by four policemen, all triple masked, wearing face visors and carrying sterilised batons. Above, there is a helicopter circling, from which the message is booming out that everyone is to clear the square. And, yes, one of the policemen has now struck Michelle and she is on the ground, and another has just stepped in and tasered her. Now, she is being sprayed all over with disinfectant, wrapped up in viral resistant plastic sheeting and is being lifted up into the helicopter, from where we expect she will be transported to the nearest Covid camp. Truly, Shane, it has been a joy to witness such police efficiency: this is what we pay our taxes for. Back to you in the studio!'

Having watched this report, I was filled with nothing but admiration and immediately looked into how I could emigrate to Australia at the earliest possible opportunity (not that, as I simply must reiterate, the Covid response in my own land has been anything but magnificent but there is something so beautifully pure in the Australian efforts that it touches the depths of my heart). Unfortunately, they are not letting anyone in at this time but, as soon as they do, you can bet that Prof. Oisín will be applying to every Australian university he can find. (I'd love to do a study on the optimal kinds of Covid quarantine camps, so if you happen to be a chancellor at a university Down Under, please hit me up).

And speaking of getting into Australia....

Aussie Covid Border Policy

The Aussies also take the right approach when it comes to keeping public health enemies no. 1, aka the unvaccinated, out. Let's take the case of the now infamous Mr. Novak Djokavic. Or should I say Mr. NoVAX Djokavic. Or perhaps even Mr. NoVAX (what a total) JOKEavic.[4]

What I admire most in the Aussie government approach is that they defended the safety of their people at ALL costs. The fact that Mr. Novax got an exemption from being vaccinated and permission to play in the tournament did not stop the Australian border force nevertheless from detaining him immediately upon his arrival and

---

[4] Ha ha, good one Oisín (Ed).

putting him into a quarantine hotel pending his deportation order. Similarly, when Mr. Novax then had the audacity to appeal this decision and win his court case, the Aussies still sent him packing, on the grounds that he was a threat to 'public health and order'. Now, that's what I call true leadership.

And they were damn right to do it.

What if, for example, some of his unvaccinated spittle lands on the ball just as he is about to serve and that very same spittle is then projected towards his opponent who, distracted at playing against someone who is so very unvaccinated, mistimes their shot and the ball ends up landing on an elderly lady in the crowd who then dies three days later because, yes, his spittle had Covid in it.

The possibility of such a tragedy, which seems eminently plausible to me, was avoided by the Aussie government's decisive actions.

Furthermore I was heartened to see just how much the Aussies wanted Mr. Novax out of their land. On the TV, I saw a young fellow of 12 or 13 interviewed: 'Well, if he does stay, I won't be watching him because he is not vaccinated', he said and with evident disgust on his face. Good on ya, son. The future is bright Down Under.

So there you go…. this is truly what a modern-day utopia looks like, a land of real-life 'Wizards of Oz' as it were. And to all those critics of Australia who say that the success of their Zero Covid approach owes a lot to their isolated geographical position….. My Lord, these kinds of people truly have left their critical faculties at the door! Do not all countries have borders? Are you telling me you can't have the armed forces at the border, ferrying all incoming arrivals off into state run camps? These kinds of things have been done before in human history and they are hardly beyond the wit of man, so don't you tell me that a Zero Covid strategy isn't feasible EVERYWHERE. Sure you could even build a wall along each border to keep people out. For anyone who has a runny nose and still chooses to travel into another country is no better than a rapist or a murderer in my view. Build a wall, keep them out and let's all keep each other SAFE.

Anyway, we have now looked at how Zero Covid can work in action and I can tell you how sad I am that we did not adopt it in Ireland (although of course I took any and every opportunity to push for it when on the radio). But that's not to say that, like in Ireland or in our next Hall of Fame country, you can't still do fabulously well. And so let us now turn to consider Canada who, while they didn't have a Zero Covid approach, did manage to achieve a level of societal backing and support that maybe, if I'm honest, leads me again to feel just a little envious...

Canada (again, a country that makes me feel a little awkward about our own Covid performance)

The Canadians have always been a wonderfully liberal nation, so different from their immediate neighbours to the south. So I was hardly surprised when they locked down the entire country and ordered everyone to stay inside. Truly, keeping track of the updates stemming from this beautiful land has been a source of nearly continuous joy. It's not all been good, of course. Those fringe truckers, mainly Confederates drafted in from Texas it would seem, were just plain grotesque...but I'll deal with them elsewhere.

I think the success of Canada's approach can be summed up by the following transcript from a TV show in which two young Canuck children were interviewed for their views on the vaccination program. The whole exchange was truly awe-inspiring: what deeply intelligent young children (and not more than 10 or 11 years old!) are being raised by our dear Canuck friends. Read on and be inspired: this is indicative of the level of society-wide support that can be achieved when your government's pandemic messaging is truly on the ball.

'Host: Are you both vaccinated?

Young girl: Yes, we both have had two doses but we are looking forward to getting more. I've asked for a Pfizer for my birthday but George wants a BioNTech.

Young boy: Yeah! That one sounds soooo cool.

Host: And are you both in favour of mandatory vaccination?

Both children: Oh like totally.

Host: What should we do with people who don't want the vaccine?

Young boy: We should call the police!

Young girl: Or maybe the army. Some of these people are extremists.

Host: And should it be mandatory for people your age as well? Should we call in the police for them?

Young boy: Definitely! Lucas is always teasing me and he's not vaccinated. I'd like him to be locked up.

Host: And how can we make people take the vaccine?

Young boy: I think we should just stick it into them.

Young girl: No, not yet. I think what the government is doing is perfect for now, cut everything from them, little by little, until they submit and get vaccinated.

Host: And why do you think people don't want to take the vaccine?

Both children: Because they are racist!

Host: Well, it looks to me like we have got some future politicians right here. Let's give these two a round of applause!

The audience cheers wildly and gives a standing ovation.'

Really, once you read this, is there anything else that needs to be said concerning Canada's approach? The government's information program around their pandemic response is so very excellent that even young children can grasp all the nuances.

Again, though, I must admit that this makes me feel a tad concerned about the approach taken in my own dear country. I mean, I know most of us were 110% behind the government, but why I didn't I see any interviews of kids like this on The Late Late Show? At the very least, Tubbers could have used the annual toy show to suggest that Santa doesn't give presents to unvaccinated kids and maybe to sing a ditty or two to illustrate the point: 'You'd better be nice, be jabbed at least twice, whether you're big or whether your small, Santa Claus will vaccinate you all!' Hmmm, better give a call to some of my mates in RTÉ and we'll see what we can come up with...

So far we've picked out the strengths of the Irish, Aussie and Canadian approaches, but, now that I truly think about it, there is one country in particular which can probably rightly be considered the King of Covid, at least when it comes to dealing with the threats posed by the anti-vaxxers. I mean the Aussies were right to indicate their disapproval of Djokavic and it's always great when you can get the kids on board, like in Canada, but sometimes you just need real action to back up your words, you know? And when it comes to putting your words into action, there is one land that, when I look back, I just wish we could all have had the courage to emulate. But, for now, their example stands as a role model for all of us, safely available for world leaders to adopt next time we are in the throes of a pandemic. And so we now move our attention on to....

Austria (ok, I have to admit that this lot really did it best)

When it comes to the cream of the Covid crop, there is no doubt in my mind but that it is our Austrian brothers and sisters who are licking it all up. I was the lead editor of a special supplement in *The Oirish Times* which, I believe, says all you need to know about Austria's approach. So make yourself a cup of tea and enjoy reminding yourself of the general Austrian geniusness on the Covid front....

'THIS IS HOW TO DO IT!

In the latest edition of this special *Oirish Times* spread examining different Covid responses worldwide, our expert-in-residence Prof. Oisín MacAmadáin shares his enthusiasm for the recent Covid restrictions in Austria. There, the government has just introduced mandatory vaccination with fines of up to €7,200 and imprisonment for those who fail to comply. Should the same happen in Ireland? Read on and judge for yourselves!

Public reaction described as "very supportive" and "bordering on the euphoric"

The mood amongst citizens on the streets of Vienna yesterday evening can only be described as one of jubilation. "Honestly, I can't remember the last time I was this happy! At last, I can safely get drunk or high with my friends, who are all vaccinated of course.... these awful delinquents have held us all hostage for far too long!" said Kirsten, a local primary school teacher. Meanwhile, Kaspar, an accountant, while also delighted at the government's decision, expressed a note of caution: "The only thing is that I feel sorry for all the other prisoners.... even the worst murderer or rapist just doesn't deserve to be near these people. Maybe they should build a new prison or camp just for the unvaccinated." "Yes, a camp!" cried another fully vaccinated reveller nearby. "We need a special camp for these people! Yippee!" Scenes of merriment and street parties continued late into the night. Incidents of public sneezing were reported but the police have confirmed that no investigations will take place. This is because experts believe that vaccinated sneezes pose no health risk to others and may even convey health benefits.

Professor of ethics describes decision as "extremely ethical" and "something Aristotle would have approved of"

Prof. Ann Schlüss of Salzburg University has said that the government decision "ticks all ethical boxes", even those of Kant whose ethical box is "notoriously hard to tick." "Look, the thing is that individual freedom must be balanced with the collective good. No one has the right to engage in casual murder on a daily basis as the unvaccinated currently are. There's no doubt in my mind but that this was as ethical a decision as you could ever hope for. In fact, I'm organising a symposium next month, the proceedings of which will be published in a book entitled: *The Jab as Moral Good: Contemporary Austrian Public Health Policy as Rooted in the Virtue Ethics Tradition*...an ideal Christmas present and yours for only €139.99!"

Loopy anti-vaxxer says he is happy to stay indoors for rest of life

Markus Nütterjob, a leading member of the 'Impfung Macht Frei!' anti-vaxxer terrorist organisation, has taken to megaphoning, from his balcony, his discontent to passers-by outside his home. "They will have to pin me to the ground!", "A fine of 7,000? Pah! Totally worth it! "Never has 'over my dead body' applied more!' A police statement released earlier today stated that they are monitoring Mr. Nütterjob's behaviour following complaints from a local resident group who believe that the embargo on the unvaccinated leaving their homes should be extended to apply to their balconies as well.

OPINION PIECE: New law reflects shift in political landscape towards Progressivism

It was only a few years ago that Austrian politics was dominated by the far-right with rhetoric mercilessly targeting Muslims and immigrants of nearly any and every nationality. The current shift, therefore, towards targeting the unvaccinated instead can arguably be seen as a death knell for the far-right Austrian movement and as a new dawn for liberal & progressive politics in the land. "I'm so glad", said local Labour politician, Hermann Hündbisket, "It was really horrible to be part of a society where discrimination against minority groups was just so blatant and every day. It makes me so proud to think that those days are gone now and that people have found a way to pick on a societal group without causing any discrimination. It's a win-win situation if ever there was one." Meanwhile, the French President, Emmanuel Macron, rang the Austrian prime minister in order to convey his congratulations concerning the recent political changes and offered words of encouragement to "keep on going emmerder-ing as many of the bastards as you can".

"We need this too!" says Cork woman on radio as poll indicates majority of public favour Austrian-style policy in Ireland

Joe Duffers' *Lifeline* program was inundated with callers yesterday voicing their support for Austria's 'Jab or Jail!' vaccination policy. "Friggin marvellous is what it is, friggin marvellous," said Trisha, a caller from Cork. "We're too pussy-footed in this country, only keeping these conspiracy theorists out of cafés and cinemas. I think the threat of a bit of time in the clink would have them rolling up their sleeves in a jiffy...I was delighted the day I got vaccinated knowing I was then fully protected but the thought that any one of these loons could still kill me just like that....so I'm all for doing what the Austrians are doing just so as to keep us all safe." Meanwhile an *Oirish Times* poll has indicated that 82% of respondents would support jail time for the unvaccinated, 13% aren't sure and the remaining 5% are currently being investigated by the Gardaí. Taoiseach Micheál Martin has suggested that a debate on a bill concerning mandatory vaccination would be a logical next step prior to its passing.'

Conclusion & Honourable Mentions

So those are my top picks for best Covid responses from around the world but please don't be disappointed if your own country didn't make the list. To be honest, even though most places haven't been as magnificent as, say, Australia or Austria, they've done a pretty darn good job all things considered. I mean, think of Panama, for example, where men were allowed out of their homes on one day and women on another or of strict-lockdown Peru where they had soldiers patrolling the streets and pointing their guns at anyone who dared come out onto their front doorstep. And, of course, dear old New Zealand and its magnificent leader, Jacinda Ardern (oh how my ardour burns for Ardern!). Now, I didn't go into New Zealand's approach in detail as it was rather similar to Australia's and so you'll have the general idea of how life under Covid has been in this other great Antipodean nation. For example, when a family of three caught Covid in Auckland's suburbs, the whole city of 1.6 million went into lockdown and even with well over 90% of adults vaccinated, the isolation period for close contacts is set at all of 24 days. Top notch stuff.

But while most countries have got with the plan, there have been a few disgraceful outliers, places that you'd be ashamed to visit, live, or, heaven forfend, have had the bad luck to be born in. Truly I hate to give attention to such countries but this is a book about exposing lunacy wherever it is to be found and so let us now turn our attention to...The Lockdown Hall of Shame!

Chapter Five: The Lockdown Hall of Shame

Sweden (or 'The Sad Story of How a Liberal Utopia Became a Far-Right Nightmare')

Oh God, Sweden, if I could have a euro for each time the loonies come out with the 'There's been no disaster in Sweden' thing, then I'd be living it up in a yacht with Klaus and all the crew. Well, let's have a look at the facts, shall we? As of the 20th July 2022, Sweden's per capita Covid death rate is a whopping... 55th in the world! Yes, you read that right. Now, how utterly terrible is that...well... somewhat awful at least.... well, actually quite respectable if you think about it....but no, no, the point is that more people died per capita in Sweden than in other Nordic countries such as Finland or Norway! So there! Afterall, that's clearly the most important comparison to make here and don't you go thinking otherwise.

Similarly, you should always be wary of people who are not risk-averse and even more so if you are talking about the mentality of a whole nation of death-seeking adrenalin junkies such as Sweden. Indeed, back at the beginning, the Imperial College Covid Deaths modelling (which you'll remember was adopted in the UK and elsewhere) suggested that Sweden's approach would lead to all of around 90,000 deaths in that country by the June of 2020. 90,000! And yet the obstinate Swedes bizarrely ignored this potential impending apocalypse and told people to go about their business largely as normal, not a mask in sight and, get this, they trusted people not to sneeze on each other! Yes, that's right, not one policeman was instructed to monitor people's sneezing behaviours. Absolute madness. And, ok, there weren't quite 90,000 deaths by that June so all I can say is that they got damn lucky but at just over 2,000 deaths well, sure, they were well on their way.

And as for when Neil Ferguson, the man leading that Imperial College modelling, came out that very June with the outlandish statement that Sweden had actually gone 'quite a long way to [achieving] the same effect (of a lockdown)'[5] all I can say is that we all have our lapses and shouldn't the man have more confidence in his own models and perhaps go to therapy to work on his self-esteem.

Furthermore, even if you were somewhat troubled by the notion that Sweden's no-lockdown approach might have been vindicated, let me remind you that you should NEVER look at things at face value.

After all, think what it is like up there in Sweden, those whole vast and empty plains, filled with roaming elk and the occasional shepherd. Sure, you'd have to go out of your way to catch the virus. Furthermore, even with those few Swedes who live in cities, they are very reserved and absolutely hate physical contact. Have you ever seen two Swedes hugging each other? Look, my point is that they are a little weird up there and these kinds of facts shouldn't be exempt from scientific discussion. Sweden is clearly a different kettle of herring altogether from the rest of the world.

So as to really drive my point home, I paste here, with her kind permission, a WhatsApp conversation I had back at the beginning of it all with my dear friend in Stockholm, Saga Loren, who was terribly concerned about the direction her country was taking and was, indeed, the first person to inform me of the nightmare that was unfolding there:

'Oisín: Hey Saga, what's up my fermented fish loving friend?

Saga: Oh Oisín, I'm soooo depressed. Everything is a total nightmare here.

Oisín: Oh, here too, Saga.... the Covid is such a beast....

Saga: No, Oisín, no.... I mean there is no lockdown here!

Oisín: Whaaaaaat?!!!!! I don't believe it!

Saga: I know. I'm like really low. People are going around as if things are totally normal....no masks or distancing...people are allowed to go into cafés and visit each other in their homes....

---

[5] As reported in *The Sun* newspaper on 3rd June, 2020.

Oisín: Oh my God! What madness…. this must be sooooo hard on your mental health, Saga….

Saga: I'm more depressed than I've ever been. It's like living in an alternate universe, seeing all of these people's faces….

Oisín: Oh Saga, I'd suggest you come to Ireland straight away but you might be an infection hazard. So you'd better stay there, I'm afraid.

Saga: Of course, Oisín, I'd never do that. I'll just stay in my studio loft for as long as it takes. I'll do the right thing even if no one else will… unlike my grandmother, dear God help us….

Oisín: What has she done?!!!

Saga: She says she is still a free being and wants her afternoon Fika, so out she goes, into the streets, says it is her risk to take…

Oisín: Here she'd be arrested for endangering public health!!!

Saga: Well, you live in a civilised country, Oisín…. but it is not just my grandmother, all the grannies are out and about and no policeman so much as even lifts a finger.

Oisín: They must be brainwashed. A perfect example of state propaganda….

Saga: Yeah, here we are lorded over by Anders Tegnell who says that lockdowns will cause more harm than good….

Oisín: That man is insane! Oh God, I'm so sorry for you Saga. Just know when the body bags are piling up in the streets that you did all you could…

Saga: Thanks Oisín, I'll try to keep my chin up and my mask on.'

I've kept in touch with Saga throughout the pandemic. It remained very hard on her but I'm glad to report that she managed to emigrate at last and has now joined me in beautiful Termonfeckin. Well, I never see her, as she is still cocooning in her flat but she says she is so much happier doing that here than she was in Stockholm. Sweden's loss is all I can say.

Obviously, it has been tragic to see how a formerly liberal and progressive land like Sweden has fallen prey to right-wing extremism, but our next Hall of Shame country has undergone no such transition. Indeed, this is more the kind of place that you'd expect would enact a doolally response to Covid, run, as it is, by a certain dictator by the name of Mr. Lukashenko who has held power there for over 30 years. Yes, indeed, I'm talking about...

Belarus (or 'The Land Where They Believe Vodka Kills Covid')

So what did this Mr. Lukashenko do when the virus threatened his borders? How did he set about protecting his people? He told them to drink vodka! This is what he said: 'I don't drink but recently I've been saying people should not only wash their hands with vodka but also poison the virus with it. You should drink the equivalent of 40-50ml of rectified spirit daily. But not at work.'

Good Lord, talk about having a laugh at the seriousness of the situation! With loony advice like that, you'd almost think that he didn't think the pandemic response in other countries was anything other than crackers. And that's what he actually implied when he also said: "I call this coronavirus nothing other than a societal psychosis, and I will never deny that, because I've gone through many situations of psychosis together with you, and we know what the results were..."

Oh, so this Mr. 'Let's Just Have a Drink and It Will All be Grand President' thinks the rest of us are all 'psychosed', is it? Doesn't he know how utterly SERIOUS this situation is?! That the world has NEVER faced a more dangerous threat?!! That our lives should change utterly and forever?!!! That human contact is obviously the most dangerous thing to our health?!!!! Would he deny that children should not wear their masks to school nor grow up in fear of dying of this deadly virus nor of killing their parents with it?!!!!! Is he honestly saying that these kinds of beliefs stem from psychosis?!!!!!! Jesus, the man needs to be sectioned.

And there are already enough alcoholics in Belarus without this kind of state blessing of the drink.

But Mr. Lukashenko, no doubt being the type who loves the sound of his own voice, had even more to say about Covid, most notably: 'It is better to die on our feet than live on our knees'. Now what kind of message is *that* for the leader of a country to give his people? Does he honestly think that risk is an inherent part of life that needs to be accepted, carpe diem, and so on? Dear God, yes, risk is a part of life but it is only there so as to be MINIMIZED and REMOVED completely. And, besides, doesn't this man know that some of us have bunions in our feet and can't stand for long periods anyway?

In any event, over the last two years, I have often thought back to humanity's previous hours of darkness and to examples of how people acted in the face of adversity. For example, my mind runs back frequently to WW2 and to all of the brave allies who fought and often gave their lives so that we now could be safe, ordering takeaways in the evening and binge-watching our favourite Netflix shows. These are the values that we in the West hold dear and which we must protect at all costs. And so what are we to make of the twisted and warped worldview of Mr. Lukashenko when, only a few months into the pandemic, he goes ahead with the May 9th Victory Day parade, commemorating the defeat of Hitler and with all of 20,000 soldiers and spectators taking part? How did he justify this egregious act? "We simply couldn't do it differently, we had no other choice. And even if we had one, we would have done everything the same. The eyes of the dead soldiers look at us, the eyes of the tortured partisans and underground fighters [...] They wanted to live but died for us." Unbelievable! And did it never occur to Mr. Lukashenko that those very eyes of the dead soldiers were probably looking on at these V-Day crowds only to see people coughing and sneezing, collapsing one by one and being taken off to the nearest morgue? And that those very eyes of the dead would probably have much preferred to live in an era when they could order chicken nuggets and chips on Deliveroo? Honestly, does this man not understand what progress really means?

And so what about the piles of corpses that have inevitably resulted from such a 'strategy'? This is where things get really sinister and just go to show how you can't trust such autocratic leaders. The supposed

Covid death toll in Belarus, as of March 25th 2022, is 6,759. That's in a country of over 9 million. In Ireland, a country with half that population, we've had almost the exact same number of deaths at around 6,693 and that despite one of the strictest lockdowns in Europe. So something seems really fishy to me here, fake news, no doubt or, at best, the civil servants responsible for tallying up the death numbers were too drunk to do their job properly.

Well, so much for Belarus, but our next country didn't exactly do the job properly either, not out of any particular malice I hasten to add, but frankly out of a certain lazy attitude and so we move on to....

Mexico (or 'Hasta Manaña Señor Covid'?)

Now, while the government approaches in Sweden and Belarus were just appalling, not every country which has had a bad innings during the pandemic has been truly atrocious. Some have just been fairly awful and Mexico falls into that category. So as to explain what I mean, I will paste here a transcript of an interview I did with a Mexican member of parliament, Mr. Manuel Tamales, at about the mid-way point of the pandemic. As you will see, I think it speaks for itself.

'Me: Mr. Manuel Tamales, you are....'

MT: My name is Manuel Tamaron, actually.

Me: Mr. Tamales, you call yourself a politician, an elected representative of the people, and yet, would it not be fair to say, that while Roma has been burning, you have simply been fiddling with yourself?

MT: I am not sure I follow your meaning, señor.

Me: I mean this deadly disease, this deadly virus, is spreading throughout your lands, and what have you done to contain it? No perpetual lockdown? You are still letting people work? And tourists in without even a negative test and unvaccinated tourists at that? And your President is after saying that the coronavirus is NOT the plague?

MT: Ah I see, Señor. Yes, we have issued appropriate advice to the people, hygiene, social distancing, and also we have a traffic light system - in areas where the virus has higher incidence, we take more proportionate measures. There is no point to lock everybody down all the time, after all people need to earn a living and….

Me: Earn a living? What?! You mean you can't just pay them to stay at home so as to keep everyone SAFE?! That's what we do here. We call it the PUP, Pandemic Unemployment Payment (and they lap it up like puppies too). Surely this should be considered a massive failure on the part of your government?

MT: Señor, we cannot afford to pay everyone in the country to do nothing. We would go bankrupt pretty quickly and it would be a disaster. Very soon, we would have no money to run anything, including the health service itself. Surely not even a rich country like yours would pursue such a mad policy?

Me: It's an essential part of the whole strategy, Mr. Tabasco. Can't you see that? Or do you think you are just too hot to handle, ha ha?

MT: Que?

Me: And your death toll is horrific! As of today (April 9th, 2021), it is 206,146! That means you have the 14th highest death rate in the world!

MT: Yes, señor, but we are a country of 126 million and our per capita death rate is not really any different from France, U.K., Poland or that hotspot country for which you yourselves introduced quarantine recently, Andorra.

Me: It's all damn lies and statistics with you, isn't it? But what I want to know is whether you have the police on the streets, arresting people if they gather in any kind of group? Are they, at least, doing their duty?

MT: Señor, just this morning, a family was abducted in my city. Yesterday, five innocent people were murdered by drug gangs for not keeping up with their protection payments. Our police are busy fighting these real issues. It should not be a crime for people to see each other. Your police really must have nothing better to do and you are lucky for that. We are doing our best here in Mexico. Life is never perfect. Good day to you.'

Well, well, well, Mr. Tortilla was really shown up there, wasn't he? Well, you might say that at least they were doing *something* to combat the old Covid. But something just isn't enough, is it? It's the whole hog or not at all, and this is really why I have singled out Mexico here so as to provide an example of the typical lack of priorities we have witnessed ACROSS the developing world during this pandemic. Why on earth is it that this lot never seem to grasp just how serious a problem Covid is? Don't they even want to become like us?

Of course, some of the more disappointing places in the last couple of years have that dubious accolade not because of the actions of their governments but because the majority of the population in question simply don't want to do what they are told. It is the Eastern European peoples that are most to blame in this case. While their governments have been dutifully ordering their populaces to stay at home and get quadrupled vaccinated, most of their peoples have been having none of it. Seemingly, though for the life of me I can't understand why, the whole Covid situation reminds them of their Communist past. Very strange. I mean, under Communism protests were prohibited, dissenters ostracised, the police controlled people's movements, free associations were banned, people lost their jobs for speaking their mind and the media touted the government line. All I can say is that you'd have to be *very* deluded to compare that to the USA, France, Australia or, indeed, Ireland of today.

Obviously, there are quite a few such former Communist countries and we need not discuss all of them. Instead, let's take one representative example and move our attention on to.....

Romania (or 'Oisín's Sincere and Heartfelt Advice to the Romanian Government')

Every morning, the missus and I settle down with our tea to read *The Oirish Times*. We take it in turns to read each article to each other and, boy, do we truly hang on every word. The whole thing has frankly become akin to a kind of religious ritual and takes us a good couple of hours. Anyway, just yesterday morning, we were both very concerned to see the following article in the Irish news section:

> 'Vaccine hesitancy in Ireland highest among Eastern Europeans
>
> A new study suggests that the highest levels of vaccine hesitancy is to be found among the Eastern European communities and in particular the Bulgarian and Romanian communities. Due to a history of government oppression, many within these groups are distrustful of government authority and conspiracy theories are common. In addition, Romania and Bulgaria have the lowest levels of Covid vaccine uptake in the European Union and.....'

'My God,' I said. 'Isn't Elena from one of those sorts of countries, dearest?'

'Why, yes, I think she is. Isn't she Romanian?'

We both looked at each other, horrified at the dawning realisation that our weekly cleaner was most likely of the view that Bill Gates was setting out to rule the world and, worse, that she was probably unvaccinated.

Simultaneously, we both said: 'We can't have unvaccinated droplets in the house.'

'But how can we ask her or convince her to take the vaccine if she hasn't?'

'I'll have to find the right words somehow.... when is she next coming? Oh, no, it is today, isn't it? Yes, it is and it is almost time!'

'Good morning Mr & Mrs MacAmadáins!', called Elena from the hallway.

My wife fled up a ladder to the safety of the attic while I positioned myself in what I felt had to be my most strategic position given the circumstances.

'Are you alright, Mr. MacAmadáin?' Elena's expression was somewhat quizzical as she spied my feet under the kitchen table.

'Oh, yes, fine, fine. It's a nice spot this. I've, um, taken to doing my research down here. It's a surprisingly good place for thinking things through.'

'Ah huh, Mr. MacAmadáin. Shall I start in the kitchen, then?'

'Ok, yes, after all we need to talk about something.'

If I were ever to admit a fault,[6] it would have to be that sometimes I don't quite find the right words for these kinds of occasions. Indeed, what happened next was all a bit of a whirl. What I do remember is that things became quite heated when I suggested, perfectly innocently, that the majority of Romanians were clearly suffering from some kind of paranoia-driven mental illness and also that Elena's last words as she slammed the front door were 'You think this isn't like what we went through under Ceausescu?! Let me tell you, Ceausescu is turning in his f**king grave that he didn't think of this! What genius to control everyone with the f**king flu! Yes, I'll get a vaccine and I'll stick it up your f**king arse!' and words of that general ilk. In short, it really didn't go that well but at least we don't have to worry anymore about the likely disastrous ramifications from unvaccinated hands cleaning our Waterford Crystal.

After the incident was over and all the broken plates and mugs had been picked up and the living room fumigated, I settled down to write to the Romanian embassy. After all, I thought to myself, I mightn't be able to convince the likes of Elena in a one-to-one conversation but perhaps I can put my expertise to the question of persuading the Romanian government itself to handle this whole situation better. My letter read:

---

[6] Please, Oisín, don't do yourself down (Ed.)

'To whom it may concern,

I am writing to offer you my advice in relation to the low uptake of Covid vaccination in your beautiful country of which I have been an avid fan ever since I watched Borat.

Having given the matter considerable thought, I believe that the best vaccination strategy you can adopt would be to make use of your traditional folkloric myths. How about a vaccination campaign centring around images of Dracula? With the text: 'Just one bite and you are immune!' Of course, you would probably have to make sure that some of the original connotations within the Dracula story, such as the idea that he had victims whom he murdered, are downplayed.

I will leave this matter to your attention but please let me know if I can be of any further assistance.

Yours sincerely,

Prof. Oisín MacAmadáin'

I haven't yet received a response although I have no doubt but that my letter is being considered by the Romanian cabinet at this very moment.

Now, while you can expect some places in the world to be not quite up to scratch in their Covid responses (not that you should make any allowances in this regard of course), there are some places where you would simply never guess this possible. And one such place is the Land of Progress, Scientific Thought and all things of that kind. But even in the US of A, all has not been everywhere well in the times of Covid, as I was to learn at my peril.....

Florida (or 'The Tale of Oisín's Nightmare Holiday')

For many years now, the missus and I have been heading off to sunny Orlando every winter to escape the Irish winter chills. This was put on hold by the ravages of the pandemic but, once we were both fully jabbed, and allowed back into the US, we thought 'why not?' to ourselves and so off we went. No problem with having a holiday in the sun, as long as everyone around you is triple vaxed, masked and keeps well away, right?

On the plane over, we were sitting across the aisle from a Floridian lady, Martha. She was nice enough, chatting away amiably. I enquired as to whether we could look forward to some nice and stringent restrictions, perhaps a night-time curfew, or to spotting unvaccinated people being rounded up in the street, you know the kind of thing that adds value to any holiday, and sure didn't what she say next turn us both a deadly shade of white: 'Oh, we don't have any restrictions in Florida. We haven't had any for well over 18 months, in fact'.

I turned to my wife, my legs suddenly like jelly and visibly shaking. 'How will I survive, how will I survive?', I muttered, trance-like. As my dear wife tried to console me, a stewardess came over to ask if I was ok. 'You have to turn the plane around', I said to her. The stewardess gave me an odd look. 'We really can't do that, sir,' before turning to my wife, 'Is your husband a nervous flyer, mam?' 'No, he's just nervous about going to Florida….', my wife replied. 'Well, he should have known that's where he was going before he booked a ticket there', and off she went, without so much as a thought for the near convulsions I was having by that stage.

I'm not one to drink but it seemed like the only option to calm down and so, a few G&Ts later, I started to drift off. My dreams were troubled: visibly sniffly men in MAGA hats embracing each other at some kind of rally about how the right to bear arms was the best defence against Covid. Little did I know that this nightmare was only a fraction of the hell on earth that yet awaited me.

As we were coming into land, I turned to my wife: 'How did we not know this? I read *The Oirish Times* every single day – I mean I write for them for crying out loud! - and not once did they so much as hint that there existed anywhere so utterly bonkers and unhinged. Shall we just stay on the plane and get the return flight back?'

'Look, dearest,' responded my wife. 'Why don't we try to make the most of it? We'll be as careful as we can and I'm sure we can still have a great holiday.'

I reluctantly agreed to give it a go and almost convinced myself that everything would work out all right.

But the next day, my nerves were not any better. In the morning, I flicked on the TV and up popped a certain Ron DeSantis, governor of Florida. 'No one will lose their job because of their personal decision not to be vaccinated', he proclaimed, 'Never on my watch and never in Florida!'

'Oh my Lord, it gets worse and worse! Anyone who serves us could be unvaccinated! Oh God…. there you go, thinking you are just buying a nice kale and oatmeal Frappuccino and instead it turns out to be a coffee of death….oh no, oh no, I can't do this! Look, honey, why don't we just stay in our hotel room for the fortnight, get room service….'

'Now, Oisín, I know you can do this. Come on, let's put on our masks and visors and go down for breakfast.'

Upon going down to the buffet, I was horrified to see a room packed full of people wandering around and with not a mask in sight. 'I'm not sure I can do this, dearest, I'm really not.' 'You can, Oisín, you can. Come, let's sit here.'

Trying to take my mind off the crowds, I picked up the local paper. A small headline caught my eye: 'Study shows 91% of Democrats are fully vaccinated while only 60% of Republicans have had at least one dose.'

'60%, 60%, oh Lord, and we are in a Republican state, that means up to 40% of the people in this room could kill me and likely even more than that as sure one dose doesn't even count…. oh Lord, oh Lord….I feel desperate….'

The next thing I knew, I had come round in our room. A doctor was peering at me with an evident look of concern.

'It would seem to be a case of acute anxiety, mam', he said. 'What this man needs is a holiday'.

'Oh no, I don't want a holiday! I want to be back in dear old Ireland, the land of the sane, I can't take it here anymore!'

'I'm afraid I can't help any further, mam.' And with that the doctor departed.

'Look, Oisín', my wife began. 'I think we need to try and do something relaxing. There was a sign for a mystery tour downstairs. I asked at the desk and they said it is totally Covid-safe. Why don't we just do that?'

How I got talked into it, I will never know, but there I sat on the bus, not realising that I was about to suffer arguably the most traumatic event of my life to date.

'Welcome to Disney Land', beamed the driver, and out we all got.

I looked around. The area wasn't too crowded. It seemed possible to keep a distance. There were physical distancing guidelines on the ground, clear directions of travel, hey, it was safe - this was going to be alright!

Just as I was starting to breathe again, I heard a voice to my left.

'Hey, wanna hug?', it said.

And before I knew what had happened, a giant Mickey Mouse had smothered me in his embrace.

Panic immediately gripped me. What were the odds that Mickey Mouse was a Republican? Very high, it would seem.... and therefore what were the odds that Mickey Mouse was.... unvaccinated?

'Um, are you a Democrat by any chance?', I stuttered.

'Why are you asking me that?', said Mickey.

'Just tell me please!'

'No, I am a Republican.'

With that, the world became to swirl, and I fainted again.

This time I awoke in a hospital bed.

'Are you sure that was the reason, Mam, that Mickey Mouse gave him a hug? It's just we've never had someone come to A&E because of that before.'

'Yes, that was it!', my voice was hoarse but I wanted to make sure they knew the truth. 'A very dangerous creature, that Mouse, he is a Republican, you know, a Republican! Why is that kind of thing allowed? He could have killed me…. he may yet for all I know….'

'This case might be beyond the scope of what we can help with, if I'm honest', said the doctor and before I could object, he was gone. My wife held my hand.

'I'm sorry, dearest,' I said. 'I recognise that this can't be much of a holiday for you.'

'It's ok, Oisín. You are right to be scared. Look, why don't we just go home?'

My eyes lit up. 'Yes, my sweet, let's just go home. Oh, I can imagine it now…. pouring a cup of tea, turning on the radio, hearing the latest case numbers and deaths, reading the newspaper about the most recent restrictions….oh, I'm calming down already! Let's do it, darling! Let's go home.'

And with that we booked ourselves on the next Aer Linctus flight back to Ireland. What a moment it was to arrive through the doors of our own home again. And not a Mickey Mouse in sight, though to this day, eight months later, I still can't watch a Disney movie with him in it without having a panic attack. My wife urges me to go to therapy to discuss my fears of him with a trained professional but I'm afraid that it will mean there is something seriously amiss going on deep in my subconscious. So, I'll just keep on going the best I can for now.

What I can say for certain is I'm never going back to such a doolally place as Florida ever again.

Anyway, in some ways I'm sorry to have had to drag you through all of the above. There are two sides to human nature and, sadly, we've witnessed its darker side in abundance over the last few years. But we have also witnessed our good side in all the areas that truly count and so I am glad now to return to sunnier uplands and to one of the most moving and beautiful aspects of these last few years…. yes, you've got it! It's time, at last, to talk about the vaccine!

CHAPTER SIX: ROLL UP YOUR SLEEVE EVERYONE!

So we really are motoring along at this stage. We've already busted some of the most pernicious myths about the Covid and also looked at the best and worst of how countries have responded to the virus. But there is so much more to come and so we now turn our attention to one of the most awe-inspiring and moving aspects of this pandemic, the vaccine. Indeed, if someone had told you, back in March 2020, that all the major drug companies would not only create a life-saving vaccine but also complete all the necessary checks and balances to ensure that it was super safe (a process that normally takes close to a decade) in less than 9 months, would you ever have believed them? And yet, that is exactly what happened! And not only that but that these major drug companies would achieve this feat using a technology that had never before been approved for any previous vaccine or any medication of any kind whatsoever for that matter...I mean, the mind boggles. These people truly are our knights in shining armour and I, for one, will be forever grateful to them.

Indeed, I'll never forget the day when I got the text from my GP: 'Oisín, a slot for your vaccination has become available'. It was the most beautiful of moments, the happiest day of my life and sure wasn't I there in a jiffy and all smiles the moment the needle went in. And then, I'll never forget the second happiest day of my life when I got the text message for my next shot. 'Oh, doctor it is so good to see you again!', I said, as he jabbed me in the arm once more. And then, oh joy of joys, came the third, fourth, fifth, sixth, seventh happiest days of my life, and each day full of more than its fair share of euphoria. And today, as I write these words, I've just come back from my eight shot and I've been assured that, at long last, I am no longer at risk of dying from Covid. Well, most likely anyway, but I'll take as many as I have to. Oh, the wonders of modern medicine is all I can say.

Indeed, I, for one, have no problem with taking the vaccine every few months for as long as I live. I am certainly not one to subscribe to the typical anti-vaxxer trope 'oh if the vaccines work, why would you even need more and more boosters, surely this shows they are ineffective'. The cheek of these people! Do they not realise they are talking about an extremely advanced new technology and that we should all just learn a good dollop of humility and patience in the face of it? So what if you need a fourth, fifth or a tenth booster? It's just a little poke in the arm, after all.

I think that Jacinda Ardern, the wonderful PM of New Zealand and fearless champion of the 'no-nonsense someone sneezes in the suburbs and the whole of Auckland locks down' approach, really framed this issue very well when she said:

> 'Your first dose is like going to kindergarten...your second dose is like going to primary school...and your third dose is like going to high school.'[7]

Damn right, Jacinda! Our immune system needs all the help it can get against Covid and it is beautiful 'natural life cycle' analogies such as these which really drive home the key message that we all need to follow this medical treatment right through to its very end.

Indeed, I'd like to take Jacinda's beautiful words and extend them even further: 'Your fourth dose is like going to university, your fifth is like when you get your first real job, your sixth is like your wedding day, your seventh is like when you earn a top promotion, your eighth is like when you retire, your ninth is like when you enter your retirement home and your tenth is like just before your funeral.'

Truly, this is the way to think about it. Do what you have to do and perform your citizen's duty.

---

[7] As from the 2.30 minute mark in her Covid update here: www.facebook.com/jacindaardern/videos/in-case-you-missed-the-details-of-our-omicron-response-package-quick-update/309617111058801/

Oh, But It is Not a Vaccine!

Far worse, of course, than the suggestion that we shouldn't all be getting our Covid shots at least four times a year from here on is the mad idea that the Covid vaccines are not even vaccines! Indeed, this is one of the most outlandish things that the anti-vaxxers come out with. Honestly, the intellectual level of these people. The scientists call them vaccines, the governments call them vaccines, it says 'vaccine' on the label, and yet still these people aren't satisfied... you can take a horse to a medical intervention designed to induce immunity but you can't force it to inoculate itself or however it is the saying goes.

'Oh, the mechanism of action has nothing in common with traditional vaccines!', the nutters chorus. 'Making your genetics make Covid, that's not a vaccine!' Look, as far as I am concerned, a vaccine is any medical intervention where you inject someone with the aim of making them immune to a specific pathogen: it doesn't matter how you get there, as long as you do.

And, oh boy, how these vaccines achieve that aim! If anything, the main Covid vaccines used to date illustrate just how utterly genius the scientists are who developed the technology behind them.[8] I mean injecting someone with an instruction to their RNA genetic system to create Covid's spike protein so that your body can then develop an immune response, now that's just awesome, isn't it? Indeed, if the story of the Covid vaccines were to be portrayed as a movie, I'd call it *The Amazing Mr. Spike* (you know, a bit like *The Talented Mr. Ripley* or *Fantastic Mr. Fox*). When you think about it, the idea of having a bit of inactivated virus in a vaccine is just so old hat, isn't it?

And as for all those anti-vaxxers going on about that Stanford study[9] which demonstrated both that the spike protein was still circulating around the body for up to two months post-vaccination (even in the

---

[8] With the exception of Robert Malone of course (a scoundrel with whom we will deal in the next chapter).

[9] 'Immune imprinting, breadth of variant recognition, and germinal center response in human SARS-CoV-2 infection and vaccination.'

lymph glands) and that the same thing does not occur post Covid infection…. honestly, they are acting as if it is bad to have the spike protein hanging around your body. Don't they realise that the more areas of your body it ends up in the better? That if it's hanging around your lymphatic system or your brain or your kidneys, that those parts are all developing their own brilliant immunity against Covid? In this way, your whole body becomes protected from the inside out and that's just one of the reasons that these vaccines are purely awe-inspiring, in my view.

Also, all this 'it's not a vaccine' nonsense frankly sounds a bit discriminatory to me. If the vaccine self-identifies as a vaccine, then who are we to say otherwise?

Mass Vaccination in the Middle of a Pandemic: Probably the Best Idea in the World

In any event, having now set the scene about just how marvellous these vaccines are, it should be obvious to anyone with half a brain that everyone needs to take them. After all no one is safe until everyone is safe. And yet the dunces in the room would nevertheless tell you that you should never mass-vaccinate in a pandemic. I know! Sure, if ever there was a time a vaccine were needed for a deadly pathogen, then surely that time would be now?! How is it that these antivaxxers honestly think that their *hoity-toity* intellectual pretences would ever fool us for even one moment?

Indeed, this most pernicious of ideas has been spread in particular by a certain 'vaccinologist' by the name of Geert Van Der Bossche. And what exactly does this Van Der Dishwasher have to say for himself? None other than that mass vaccination while a pandemic is raging will only lead Covid to become all the smarter as it detects widespread vaccine-induced antibodies and then rejigs itself to create new and more infectious variants. In other words, the vaccine program is merely a catalyst for driving many new iterations of Covid which, ultimately, the current vaccines won't be much use for. Honestly, can you believe this man?! While I 100% agree that Covid is extremely smart (unlike this total chancer), there is an obvious flaw in his argument. If the mass vaccinations fuel new variants, then all you need to do is to create new vaccines! There, problem solved. Your current wash cycle is well and truly over, Geert!

Anyway, people such as Geert clearly only have first world problems on their mind and you can be sure they never concern themselves with our less fortunate fellow humans. I mean, what about the developing world? Should we deny people there access to the vaccines just because people like Geert don't think anyone should be needled at all? Of course not! Indeed, let me tell you a little story, one that will hopefully lift your spirits and inspire you as to what you can really achieve if you put your mind to it....

Getting the Vaccines to Those Who Need Them Most

As unbridled a success as the vaccine rollout has been in my own land and so many other developed countries, sadly this hasn't been the case everywhere in the world. For someone such as myself who so wishes for the benefits of science to be spread as widely as possible, this has been the source of the greatest sadness. 'Why are you crying, Oisín?' my wife often asks me when I awaken in the middle of the night, blubbing away to myself. 'It's because there are so many places in the world that haven't got the vaccines yet dearest!', I will reply. 'You know places in the developing world like Rwanda or El Salvador'. She then puts her arms around me and we cry ourselves back to sleep.

But on one such occasion, my wife had a bit of a eureka moment.

'Oisín, do you remember that place we saw on the TV the other night, the little mountainous republic in Europe that we'd both never heard of... what was it called again?'

'Um, um, oh, yes, the F.S.R.B., The Former Syldavian Republic of Bogrenia, wasn't it? It's amazing the places that exist which you've never heard of, isn't it?'

'Absolutely, darling. Well, I'm just googling them now, and sure it says here that their current Covid vaccination rate is only 0.3%. Why don't you help them out? You know, start a charity, 'Jabs for

Bogrenia' or something like that? And then you can head off there and set about vaccinating the whole nation? It says here that the population is only 23,000. To think, if you could protect everyone there, what a difference that would make. Not everyone can say they've saved a whole land, you know…'

'No, I couldn't, dearest, sure, what qualifications would I have to do something like that….'

'But you're an expert, Oisín!'

'Oh, so I am! Well, I suppose…. yes, maybe I could! No…I will, I will!'

And so six months after this fateful conversation, dearest reader, I found myself on a plane to Brámstokeravia, the capital of the F.S.R.B. It was, I must admit, a slightly awkward flight as I was the only passenger (I had had to book all of the seats in order to take 150 bags of life-saving vaccines in the hold) and the flight attendants looked at me as if I were the strangest man alive. But I just kept smiling back at them and assuring them that I had plans to save their entire nation. Well, I'm not sure they saw my smile through my three masks but I did my best to convey my good intentions.

Having hired around 40 taxis at the airport, I arrived at the hotel to find that only 10 or so of them had followed me and the rest had taken off with the vaccines. 'Oh, well, no matter,' I thought to myself, 'I'm sure they will still give the vaccines to their friends and family to take…. what does it matter whether I administer the jabs or they do? And I still have loads of them, enough for my clinic, for sure.' I then went for a much-needed night's sleep.

The next morning, sweating and panting after fighting my way through all the luggage in my room, I went down searching for my usual vegan breakfast of chia seeds on a bed of kale. Having found this to be unavailable, I poured myself a cup of coffee and picked up the local paper. It's always good to learn a bit about what is going on in the countries you visit…however, I must admit that the headline took me aback somewhat:

"DRUG BARON VLADIMIR THE LACERATOR STAGES SUCCESS-FUL COUP AND NOW CLAIMS DOMINANCE OVER THE CITY FOLLOWING USE OF POWERFUL, NEW WEAPON

Brámstokeravia is now under near total domination by the members of local drug baron, Vladimir the Lacerator following their successful coup yesterday evening. The sudden turnaround in the political fortunes of the country seems to be thanks to the devastating use of a new weapon which left both government and its military leaders quaking in their boots and with no option but to hand over power.

Yesterday evening, hundreds of Vladimir the Lacerator's gang members stormed the parliament carrying what appeared to be syringes. The politicians and soldiers merely laughed at them until one unsuspecting political aide was injected by one of Vladimir the Lacerator's followers and promptly dropped dead. Whereupon, out of fear and intimidation, the Prime Minister resigned and handed over all power to Vladimir the Lacerator

IMAGE OF PRESIDENT SIGNING OVER POWER WITH VLADIMIR THE LACERATOR NEXT TO HIM HOLDING SYRINGE TO HIS NECK...."

At this point, I looked out the window to see soldiers patrolling the streets 'armed' with the very vaccines I had brought into the country only yesterday. Two such soldiers merely had to point them in the direction of some youths loitering on a street corner for the latter to take off home, crying for their mummies.

I must admit that, at this point, I was feeling a little sheepish. But I tried to remain optimistic. After all, the aide's death was almost certainly an unfortunate coincidence and nothing to do with the vaccines (which are safe and effective) and if the new government happened to be using the vaccines as its primary tool of law and order, well, surely this would actually end up bringing a lot of benefit to the country. So, as I finished my morning coffee, I actually reflected that, on balance, my trip had so far brought much more good than harm to this fine nation.

Greatly buoyed up by this fact, I set off to find the building which I had acquired prior to my arrival and which I hoped to use as my vaccination clinic. It was on the outskirts of the city, a fine location, poplar and spruce trees all around, and plenty of fresh air. Being a somewhat more rural location, it was also surrounded by goats, happily munching away on any and everything. Indeed, I had read in my guidebook that goats outnumbered people in this wonderful land by at least five to one.

'This will do very well', I thought to myself, 'sure I might as well get started'. So, taking one bag of vaccines, I set up a table and put up a small sign: 'Free vaccines for all'. Then, I sat down and waited.

Nothing much happened for a while. But then a middle-aged man, with a goat on a leash, approached me.

'You are here to vaccinate my goat?', he enquired.

'Um, no, these vaccines are just for humans...'

'But my goat is sick and you say 'vaccines for all'.... please vaccinate my goat now and make him well.'

Although technically untested in goats, I felt that the benefits would probably outweigh the risks for this particular goat and might even help it if any rather capricious (geddit, geddit?)[10] and goat-infecting strain of Covid were to emerge in the future. And it would at least get the ball rolling on my own efforts to vaccinate the nation.

Within seconds of his vaccination, the goat started to froth at the mouth and flopped over onto his back, flailing his legs in the air, and making the most terrifying noises.

The man looked at me disapprovingly. 'Your medicine is not making my goat any better', he said rather, even I would have to admit, astutely.

'Well the E.U.A. doesn't extend to goats....', I muttered, before adding 'but look I'm sure he will be better in a moment. See, he has stopped bleating now...'

'That is because he is dead.'

'Ah, so he is.'

---

[10] Honestly, Oisín, you are such a card! (Ed.)

'You have killed my favourite goat.'

'I'm so very sorry, is there anything I can do to make it up to....'

'You have killed my favourite goat! Everybody, come here, this man has killed my favourite goat!'

I suddenly found myself surrounded by a gang of around 20 rather strong looking men who seemed rather too intent on avenging the unfortunate fate of the goat, a fate which no doubt they feared I intended to inflict on their own goats and which I, quite frankly, feared they were about to inflict on me....

Just as one of them was about to strike me with what appeared to be a scimitar (though I can't be absolutely certain on this point, to be honest, my eyes were decidedly closed at that moment), a voice called out from beyond the crowd.

'Stop right there! I wish to speak with this man.'

'Yes, boss.'

The crowd gave way and I found myself face to face with a man wearing all black leather, ammunition straps and with an AK-47-armed bodyguard on either side.

'My name is Drakulblüd, and I am head of the Drakul gang. Where did you get these syringes and are they the same ones that are being used by my sworn enemy, Vladimir the Lacerator, a man who has now so shamelessly taken over our country?'

Spotting a way out of my less-than-ideal predicament, I avowed that yes that was indeed the case and, dear God, if he wanted all of them, he could have them, and whatever else came into my head – you know, the kinds of things you say in such situations.

'I will spare your life in return for these syringes. Take me to them.'

Later that day, the vaccines having been handed over, I decided it was time for me to take my leave of The Former Syldavian Republic of Bogrenia. During the taxi ride to the airport, I spotted whole gangs in the streets, injecting each other with syringes and, in the distance, clouds of smoke arising from the government buildings. In the airport, reflecting on my trip, I concluded that, while things hadn't quite worked out as I had planned, the net result was that countless thousands of the people of the F.S.R.B. would now end up vaccinated against Covid and that this, if anything, was a situation where the end surely justified the means.

So I patted myself on the back for a job well done and looked forward to getting back to my wife in dear old Termonfeckin and telling her all about it.

Vaccinating Our Furry Friends

I can only hope that the tales of my successful trip to The Former Syldavian Republic of Bogrenia will inspire governments and NGO groups everywhere to send vaccines to the countries where they are most needed. After all, no one is safe until everyone is safe. Indeed, while I was pondering the undeniable truths inherent in this fact the other day, I suddenly realised, much to my horror, that we have not quite been spreading the net widely enough. For, and I hate to say this, has our approach to the vaccine roll out not been a tad... species-ist? I mean, isn't it the case that some animals are also catching the Covid? Why haven't we developed a vaccine for *them*? And not just for their sake but also for our own as isn't there a strong likelihood that transmission among animals could lead to truly monstrous new variants, for which we will then need new vaccines, and so on and so forth? Therefore, it was with quite a start that I realised the *actual* truth of the situation: no one is safe until every human, cat, dog, bat, anteater, hamster, kangaroo, pangolin and, indeed every single one of our furry friends, is double vaccinated and with a booster program in place for the rest of *their* lives.

Now, critics might say that this is an unrealistic and truly mammoth task (well, they at least don't have to be vaccinated!). But while, yes, this is about the safety of *all* of us (for the reasons just mentioned), it is also about the welfare of our pets who, so far, have not had a voice, or even so much as a squeak, throughout this pandemic. All along, our cats and hamsters have been suffering from the worst sniffles and colds of their little lives, and there has been no one to speak up for them. And, worse, some of our pets have no doubt ended up with 'Long Covid'. My cat, Fauci, for example, spends all of his time lying around doing nothing at all. I can only imagine that this is because he is suffering from the dire effects of this devastating post-viral illness. Similarly, I am also deeply concerned about the sneezing fits I see in Klaus, my paraquet, and I do fear the distinct possibility of a mutated form of Covid amongst our feathered friends. This would probably be called the 'Avian Transmitted Covid Plague of Death Disease' (ATCPDD), or some such. You heard it here first, anyway. So, in short, if we care about reducing suffering wherever it is to be found, then we simply have no choice but to start designing vaccines appropriate for each and every species on the planet.

I cannot emphasise enough, however, that it is also strongly in our own interest to take this course of action. I alluded above to the possibility of new, 'monstrous' variants emerging as a result of animal transmission and I do not use that word lightly. Indeed, we may end up facing 'a whole other kind of beast', as it were. Personally, I don't think it is at all far-fetched to imagine a scenario in which a more virulent strain might even lead our pets to rise up and take to the streets, nibbling with their Covid-infected teeth at the unsuspecting ankles of humans passing by. Well, we should not rule it out anyway: no one wants a rebellion of runny-nosed and coughing hamsters.

And, if this kind of thing were to come to pass, we would be faced with absolutely no choice but to instigate mass cullings of all unvaccinated pets and animals. There is no doubt in my mind but that it was this threat that the Danish government were wise to when a handful of mink caught the Covid there. For the Danes acted decisively, killing not only the Covid-infected mink but also every single mink in the land, around 17 million of them, all told. Who knows what those mink would have got up to if they have been given half the chance.

So, my message is clear: if we don't get around to vaccinating all animals everywhere now in order to save their (and our own) lives, there may come a time, not long from now, when we will have to kill nearly every single non-human creature on this earth. So, frankly, we better get on with a new animal vaccine program pronto.

And once the cat and paraquet vaccines are available, you can be sure that Fauci and Klaus will be first in line.

Well, it's been a delight to ponder the arrival of these miraculous vaccines in this chapter. Bizarrely, of course, not everyone is of the same view as you or me, dear reader. Indeed, there is a certain group amongst us who not only do not see the benefit of these vaccines for animals, the developing world or even for themselves for that matter...yes, at long last, I am talking about the anti-vaxxers! The main villains have finally entered our story. And so, let's show them who is boss once and for all, eh?

CHAPTER SEVEN: ENTER THE ANTI-VAXXERS!

Up until this point, we have busted plenty of the myths that are put out there by the Covid-denying nutcases that are among us but now, at long last, we come to the most despicable myths of all, those that center around the vaccine. For people who spread misinformation about the vaccine are no worse than murderers in my view.

And that's why both this and the next chapter are arguably the most important in this book. Read them very carefully so that you are armed to counteract the anti-vaxxer lies once and for all.

But where shall we start? I suggest that we first bring our attention to a couple of the anti-vaxxer 'ringleaders', as it were, the people to whom the conspiracy theorists look for guidance and inspiration. Expose the those at the top and whole edifice crumbles....and there are two in particular whose pronouncements abound in the loopier corners of the internet, men by the name of Robert Malone and Peter McCollough.

So, to begin with, let's put these shady characters into their proper places, shall we?

Robert Malone: The Greatest Antivaxxer of Them All

You'd be hard pressed to come up with someone who plays more into the antivaxxer narrative than Dr. Robert Malone. And why? As, apparently, even though Dr. Malone 'claims' to have invented the technology for the m-RNA vaccines, he nevertheless casts doubt on their safety and is a prominent critic of the vaccine rollout. If the inventor of the vaccine tech has safety concerns about the vaccines, shouldn't we all listen to him, blah, blah, blah, kind of argument, honestly you couldn't make it up even if you tried.

Anyway, what a load of baloney. Indeed, there are four main reasons why I wouldn't trust a single word that comes out of Malone's mouth.

First of all, just take one look at him and you'll immediately notice that he has a big beard. Now, this is a trait shared by many anti-vaxxers: they live in caravans in the back of beyond, wander around in their underwear scratching themselves and generally become very lax in matters of personal hygiene. Beards are a natural consequence of this and, as such, they are a *very* bad sign.

Secondly, so the man 'claims' to have invented the technology for the m-RNA vaccines, does he? Says he holds the patents for this technology? Well, I very much doubt this. Why on earth would someone who has invented a technology which is now being used to SAVE THE WHOLE WORLD not want to take ALL the credit that is due to him? Why would such a person risk being pilloried, slandered, ostracised by the mainstream media purely on a matter of supposed 'principle'? The whole notion strikes me as utterly outlandish...I don't believe it and neither should you.

Thirdly, the man owns a farm and apparently he has horses on it. This makes me immediately suspicious that he's probably ordered loads of the horse deworming pill, Ivermectin, and no doubt has some with his morning coffee as part of some quack Covid prevention protocol. I can picture the scene now.... in his camper van, his wife, whose name is probably Betsy or Bugsy or some such, calls over to him: 'Bobby, dear, do you want one or two spoons of horse-dewormer in your coffee?' 'Two, my dearest, and don't forget to add a pinch of bleach.'

Fourthly, he was removed from Twitter for posting misleading information. If you ever needed proof that the man is spouting anti-scientific drivel, then you need look no further than this fact. Twitter upholds the highest standards of scientific discourse on their platform. I mean, remember my point from earlier that I'm fairly sure their factcheckers must probably have, at minimum, PhDs in virology... these guys are so smart that can spot the minutest scientific error that wouldn't cause the rest of us even to bat an earwig, as the old saying goes. Therefore, if Twitter feels that the man who supposedly 'invented' the m-RNA tech is voicing misguided and harmful concerns about that tech's use, then I have absolutely no reason to doubt them.

How is it that anyone can take this man seriously? It really does beggar belief. This pandemic has, indeed, been a pandemic of misinformation as much as anything else, has it not?

This phenomenon is also in strong evidence as we turn now to Peter McCollough, another of the most prominent anti-vaxxers out there.

Anti-Vaxxer Ringleader No 2: Peter McCollough

So who is this man, exactly? Well, if you were to believe him anyway, he's a leading US cardiologist, with no less than the most published papers in his field, and someone who testified to the US senate about early outpatient treatment for Covid in November 2020. Honestly, is this not just the kind of insidious stuff we have come to expect from anti-vaxxers? Can you see how cleverly they can dress themselves up to appear just like experts? Why is it that these people don't realise that experts only exist on our side of the fence?

To be frank, I'd already had enough of this charlatan just from reading his bio but, in the interests of keeping the wider public wise to Covid misinformation, I forced myself to research him just a bit more. And what I found is that, of course, he appeared on none other than the Joe Rogan show, arguably the world's most influential podcast and one that has featured its fair share of Covid-deniers. In forcing myself to listen to Rogan's interview of McCollough, what I found was one very slippery customer indeed. For example, McCollough had the audacity to claim that he was primarily interested in *saving* lives from Covid-19 (I know, you couldn't make it up!) using what he called 'early treatment protocols'.....early treatment, what a laugh, we all know the only scientific way to deal with Covid is to stay inside for two years waiting for the vaccine, take seven of them, and then stay inside some more. And yet there he was unashamedly suggesting that hundreds of thousands of lives could have been saved if a different approach had been taken... God, how sick, I thought to myself, that an anti-vaxxer should claim to care about saving lives from Covid. I simply couldn't take it anymore and so I stopped listening right then and there.

So as to save you from having to endure the same fate, here is a transcript of how I imagine the rest of his conversation with Joe Rogan went, once McCollough dropped his saintly mask:

'Joe: So, tell me, is Bill Gates out to get your hamster?

McCollough: Absolutely. And he is not only out to get my hamster but yours and everyone else's. In fact, I've had to put Hubert and his cage in an unidentified location so that Bill can't find him.

Joe: That seems like a sensible precaution to me. I might do the same with Harald. So, really, this is a worldwide effort by a group of elites who have a depopulation agenda aimed at hamsters?

McCollough: Yes, that is precisely what is going on. It's well documented. The elites have long since identified that a hamster depopulation program is a potentially lucrative exercise and that's what this is all about. The vaccination program for humans is only a sequel to the real agenda which is to get rid of all hamsters.

Joe: That makes total sense. But why not go straight to vaccinating the hamsters, why a program for humans first?

McCollough: A worldwide hamster vaccination program wouldn't make much sense to people on its own and out of the blue. It would seem, shall we say, a little odd and people mightn't accept it. Far better to get everyone used to the idea that there is a deadly virus and that we need to vaccinate all humans. Then, if the virus happens to 'spread' to hamsters and they become a 'threat'....

Joe: Then the hamster vaccination program becomes an obvious necessity?

McCollough: Exactly.'

Now *that*, folks, is the kind of ludicrous carry on that anti-vaxxers really believe and don't you let them or anyone else convince you otherwise. Early treatment protocols... jeez Louise, what absolutely rubbish.

But while we are on this topic, I feel that it behoves me to comment on the great Joe Rogan Spotify debacle in general…. sure, Joe really didn't know what he had coming for him when he decided to interview some of the leading tinfoil hat conspiracy theorists, did he? Not only did several world-famous artists withdraw all of their music from the platform as a result of these podcasts, but Prince Harry and Meghan also stood up, as is their usual manner, for all that is beautiful and true and expressed their sincerest concerns as well. It's amazing how stupid these conspiracy theorists are when, even with all their supposed medical degrees and years of research, their level of knowledge is such that they can still easily be shown up by the likes of a Neil Young or a Meghan Markle and others who don't have any training in virology or vaccine development AT ALL. Now, what does that tell you about just how thick these conspiracy theorist leaders really are?

In any event, we have now exposed two of the leading anti-vaxxers which is enough to give you an idea of the kind of characters people will be duped by.

But what about the negative impact which stems from the misinformation that the likes of McCollough and Malone spread? Indeed, it would be all very well if the anti-vaxxers were just talking among themselves, keeping to their own paranoid little bubble, but their dangerous misinformation has real world consequences in that it leads ordinary, decent people to suffer from a terrible, new illness, an illness which I shall now describe in detail….

Vaccine Hesitancy Disease

Now, an old friend of mine, who happens to be a doctor, came round the other day for an appropriately socially distanced cup of tea (we're both fully vaccinated many times over but you can never be too careful). He stayed out in the garden, I opened the kitchen window, and we both shouted at each other through our masks. It was a little hard to make out everything he said what with the wintry gale that was blowing but I actually managed to understand him pretty well. Basically, he informed me that he is increasingly diagnosing more and more of his patients as suffering from a terrible new illness called 'Vaccine Hesitancy'.

'Some of them are even in the terminal stages of the disease!', he shouted.

'Oh, really? Give me an example!', I cried.

"Well, one long-standing patient, an elderly lady called Margaret, came to see me a few weeks ago. 'Doctor', she said, 'I'm very hesitant about taking the vaccine.' (And so, right there and then, I had my clear diagnosis.) 'And why is that, Margaret?' 'Well, I have a neighbour who took the vaccine and then, when he visited me the next day, he had a stroke. His speech was very slurred, but I think his last words were "Whatever you do, don't take the vaccine." 'Ah, I see, Margaret, you see that kind of person is what we call an "Anti-Vaxxer"'. 'Oh, I see, doctor. I didn't know that. You mean they can seem to be like ordinary people, people you've always thought were just normal and good?' 'Absolutely, Margaret. It can be quite a shock when you find out what some people are really like.'"

I paused my friend at this point. 'She didn't actually think her neighbour's death had something do with the vaccine, did she?' 'She did, indeed.' 'But does she not know that correlation is not causation, post hoc is not propter hoc and all that?' 'Well, that was the point I made to her next....'

And so, my friend continued his story:

"'Well, yes, Doctor, I had no idea Séamus was an anti-vaxxer. He always seemed so normal. But thank you for your reassurance. However, there's something else.... I saw this thing shared on Facebook which said there have been over 28,000 reported deaths in the US on something called VAERS and that normally a vaccine is pulled for investigation if there are 50 deaths....' 'Now, Margaret, let me reassure you on this too. You see, there is a principle in medicine which states that 'correlation is not causation' .... just because someone dies shortly after the vaccine doesn't mean it was the actual *cause* of their death. They might just have tripped over their cat or something - there is simply no way of knowing.' 'Ah, I see. So all those deaths might.... or then again might *not* have been caused by the vaccine?' 'That's right.' 'Well, that is somewhat reassuring... I suppose.' 'Good stuff, Margaret. So would you like a jab now? I have loads here.' 'Um, well, hang on, Doctor, there is one other thing...'"

'My God, she was STILL suffering from Vaccine Hesitancy even after all your reassurances?' 'Yes, indeed. As I said, it was a near terminal case of the ol' hesitancy, but I managed to turn it around in the end, Oisín.' 'And how did you do that?'

My friend went on:

" 'Yes, Margaret: what else concerns you?' 'Well, I did some research online myself and came across a paper from the University of Stockholm which suggests these vaccines inhibit DNA repair in vitro and that this could have very serious long-term health implications.[11] We all know that DNA repair inhibition can encourage cancer and so forth....' 'Now, Margaret, Margaret, let me stop you right there. Surely you know that you can't trust every old thing you read on the internet? Sure, why do you think it is that we encourage people never to Google anything about their health? It's just for these kinds of scenarios, so that you don't get unnecessarily scared.' 'Oh, well, I see, Doctor. So in your expert opinion, this study is nothing to worry about?' 'Of course, not Margaret, of course not. For starters you are very old, and it takes cancer ages to develop... so look, how about getting your first jab today then?' 'Oh, ok, Doctor, you have convinced me!' 'Good girl, Margaret, good girl. Now, here we go, just a little poke....and done!'

'Oh well done!', I congratulated my friend. 'You managed to cure her of her illness in the end!'

And while my congratulations were genuine, my friend's account still troubled me greatly. Indeed, it only goes to highlight the nefarious nature of these anti-vaxxers, whether her neighbour's blatant spreading of misinformation at the moment of his death (typical emotional manipulation, if you ask me) or presenting adverse event data without expert interpretation. No wonder Margaret then ended up doing some hypochondriacal Googling of her own. To be honest, the whole conversation with my friend really opened my eyes about the extent of the misinformation threat we are facing (and it also left me hoarse from all that shouting).

---

[11] I refute this outlandish paper in the next chapter.

As my friend was leaving, I asked him how Margaret was doing now.

'Ah she is just in hospital at the moment recovering from a major stroke but at least it wasn't the Covid that landed her there.'

The Anti-Vaxxer 'Freedom Fighters'

Well, we have just seen the damaging effects the antivaxxers can have on perfectly innocent people but that's not everything of a nefarious nature that they can get up to. Indeed, they are also increasingly having the gumption to take to the streets and cause societal unrest, protesting for ideals such as 'freedom', 'bodily autonomy' and other such ideas which only go to show just how mentally ill they all are.

Worse still is that these gatherings take the valiant members of our police force away from their general duties which are hard enough, such as tackling incidents of sneezing in public, and put them on the Covid 'front lines' as it were. How brave are our policemen and women in facing down the far-right spluttering of these undesirables! I remember well the scenes of chaos in my own fair city of Dublin when the protestors filled the streets there. I'm glad to say, however, that they were no match at all for the young men and women of An Garda Síochána who were all triple masked, wearing protective visors and carrying sterilised batons.

But much as I admire Dublin's wonderful police force, I'm once again forced to admit, albeit it with the greatest of reluctance, that other countries have pipped it to us on the Covid law enforcement front. Indeed, when I heard of dear President Macron's announcement that he wished to 'piss all over' the unvaccinated 'until the very end', I knew that I needed to see for myself how a State with the military might of a country such as France tackled the antivaxxer protests. Simultaneously, I also sensed that the merde was very much about to hit the fan on the streets of Paris and that I needed to be there to cover it. And so I rang my editor at *The Oirish Times*, he gave the go ahead and, sure, off I went. A few days later, this appeared in the front pages….one of the better pieces of my many excellent articles, I must say:

"VERY BRAVE FRENCH SOLDIERS IN TANKS AND ARMED GENDARMES FACE DOWN EXTREME, POTENTIALLY LETHAL THREAT OF ANTIVAXXER DROPLETS

When tens of thousands of cars and trucks belonging to the so-called 'Freedom Convoy' set out from all around the country to invade the centre of Paris, President Macron decided that enough really was enough. In a televised address to the nation, the President set out his plan: 'We are being descended upon by an army of extremists, anyone of whom could be carrying potentially lethal amounts of Covid. So I will call in the army to emmerder them.'

The reference to 'emmerder' is representative of Macron's preferred health policy for dealing with the unvaccinated. Typically translated as 'to piss on' in the foreign press, I'm delighted to report that its true meaning is actually 'to shit all over' (ah, is not French the crème de la crème of languages everywhere?) The President's announcement brought widespread relief to Parisians who were understandably afraid that Paris could become yet another Ottawa. 'Oh mon Dieu', said one resident, 'my wife and I were so afraid there would be a lot of 'onking through ze night and zat zis would interrupt our love making.'

If the antivax protestors were expecting to take up residence in their cars and trucks on the Champs-Élysées, they were to be sorely disappointed. On the outskirts of Paris, they were met with tanks, specially deployed for the occasion, and fully armed soldiers and Gendarmes. Convoy cars were stopped by Gendarmes pointing their guns at the drivers while the tanks halted the protestors in their tracks. Parisian head of police, Michel Moustache, is reported as saying: 'We were very concerned, to be honest, as we were not sure that our tanks and guns would be a match for the large amounts of unvaccinated spittle that could have been issued in our direction. But, in the end, we managed to turn the convoy around pas de problème, just in time for us all to have a nice four course lunch culminating in a marvellous crème brulée and washed down with some Burgundy.'

Oisín MacAmadáin is an expert in residence at *The Oirish Times*"

The French don't mess around, let me tell you! I think we could learn from their approach in Ireland. Of course, for that, our army would actually need to have a tank…. but if the threat of antivaxxers marauding through the streets isn't enough for our government to prioritise the military budget a tad more, then I don't know what is.

I should, of course, at this point mention the pesky, fringe truckers in Canada that started this whole 'freedom convoy' business in the first place and, as it so happens, I decided to give them their own section. And so…. read on!

Oisín Heads to the Far North: Meeting the Far-Right, Anti-Vax, Canuck Truckers

I'm not one to name drop but I'm actually best buddies with the Canadian PM. Not only do his devilishly good looks make me swoon, much to my wife's concern, but I have also found his whole Covid policy to be…. well…. how should I put it: if a Covid policy could be orgasmic, then his would be it.

So when I heard that a whole army of truckers were descending on the Canadian capital to protest my buddy's efforts to save lives, I immediately packed my ear muffs and booked my ticket to Canada: I wasn't going to let my dear friend face this army of virus-denying nutters alone.

But before I set out, I gave him a call so as to get the lowdown on the situation.

'So what is going on, my Trudy-wudy?', I asked.

'Oh, I so love it when you call me by that name, Oisín…. anyway, basically these people are keeping society hostage. They are forcing shops and businesses to close, ruining the livelihoods of so many of our citizens, and they are keeping people confined to their homes for fear of their safety should they venture outside. Truly, they are despicable!'

'Oh, they are! How dare they do these things, Trudy-wudy?!! I mean, you'd never do any of those things to anyone, would you?'

'Of course, not, Oisín-woisín.'

'So who are these people?'

'Well, they are like totally fringe.'

'Ah, huh'

'And like totally racist, of course...'

'Of course. Probably going around all black-faced up and the like, uhhh!'

'Um, well, yes, maybe.... but they are also like totally misogynistic....'

'No surprises there. Anything else?'

'They hold like totally unacceptable views.'

'Got it....is that all?'

'Well, they are both like totally white and totally male....'

'That goes without saying, Trudy. Basically what you are saying is that whatever you are, they are not, and vice versa?'

'Absolutely, Oisín. I would rather die than be a white man...oh, and it seems that most of them aren't even truckers. They are mainly far-right activists who have just been drafted in, most of them probably from Texas. They were even spotted waving posters with swastikas.'

'Ah, thanks, Trudy, you've given me a such a clear description. I'll be with you in a jiffy and I so look forward to seeing you!'

'Oh, me too - see ya soon, Oisín!'

Anyway, next thing I know, I arrive in Ottawa. I gave my buddy a call again but couldn't get through and found myself redirected to his secretary. 'I'm afraid the PM is unwell, Prof. MacAmadáin, he has Covid and needs to isolate.' The full gravity of the situation then dawned on me: this army of racist, women hating creeps were having a go at my dear Trudy and at a time when he was no doubt facing the ravages of this most dangerous disease. For all I knew, he could have been at death's door and the last thing he would ever hear would be the blaring of horns mixed with regular expletives directed at him.... And so I became determined to face up to this 'army' of truckers myself and get them to turn their convoy around.

I took a taxi to parliament square and there they all were, making an awful ruckus and protesting away....

I went over to one of the trucks. 'Oi, you! Yes, I said, you! I want to talk to you!'

Out got a woman with brown skin, long black hair and a certain majestic poise.

'Um', I said, 'did they capture you or something?'

'I'm sorry?'

'It's just that I wanted to talk to one of the truckers who are causing all this fuss and, well, you wouldn't appear to be...'

'I am one of the truckers who, as you put it, are causing all this fuss. What do you want to talk about?'

'And you are Canadian, then?'

'How dare you? Are you referring to the colour of my skin? I am First Nations and, yes, I am Canadian.'

'I see'. I looked down at my notes and struggled to reconcile them with this latest turn of events.

'Hey, Marty', the woman called out. 'Come over here – I need your help!'

'Yes, Nagamo, I'm coming right away.'

I was now faced with a burly man who was, overweight, bearded and, thanks to be God, undeniably white. 'Phew', I thought to myself. 'Now, I am on steadier ground.'

Just as I was about to confront this man about his evident racism, Nagamo spoke:

'Marty, this man is a racist.'

'Oh, what...?!', I stammered. 'No, I'm not...'

'Yes, he is. He couldn't believe that I'm a Canadian.'

'Oh no!', said Marty. 'I don't believe it. We don't go in for that kind of thing in Canada.'

'I'm so sorry', I said, 'It's just I was told that you were all racist….'

'But you are the one who was so keen to point out the colour of my skin. And who, exactly, told you that?'

'Um, I really can't say, sorry….'

At this point, I felt I needed to get back on the offensive with some kind of criticism I could be surer of….

'Well, Marty, it would seem pretty damn clear to me that you are *not* a trucker!'

'I most certainly am. And have been all my working life.'

'You mean to say that that truck right there is your own? I very much doubt it!'

'Ah, I see, you believe that we aren't really truckers, that we've stolen all these trucks and that the real truckers are bound and gagged somewhere. Or perhaps that we all just happened to have a spare truck in our backyards, as you do, and when the opportunity came for us all to dillydally in a spot of right-wing fascism we popped into our trucks and set off…. well, let me tell you, those are all conspiracy theories.'

'Uhhh, no, that can't be the case, you are the conspiracy theorists, not us!'

'Yeah right, and you are the ones who think it makes sense that the world should stand still because of a virus with a less than 0.1% mortality rate…and you think we believe crazy things!'

I was feeling increasingly unsure of myself, such was the twisted sophistry I was encountering, and so I went for my final trump card, the one thing I knew for certain would show this man up once and for all.

'I bet you have a swastika though!'

'I do not.'

'Yes, you do!'

'No, I don't. That whole swastika thing was someone waving a poster saying that a swastika was representative of the kind of government we have now. That the media should then use that against us only goes to prove the point....'

'No, that can't be true! My friend Trudy-wudy said so!'

'Trudy-wudy? I'm sorry, that's not who I think it is, is it? Wait, are you friends with...., wait, mister, wait!'

I must admit, dear reader, that, having blown my cover in this way, I felt it was probably best to get the hell out of Dodge. And so I quickly found myself back in the safety of the airport where antivaxxers are naturally not allowed. Reflecting on my conversation with Nagamo and Marty, I concluded that they probably weren't real truckers, but some kind of paid actors put in by white supremacist groups to fool people into thinking that normal, liberal sounding Canadians were part of the trucker convoy. I'd just had rotten luck, in other words. So I put the whole incident out of my mind, knowing that I had, at least, done my very best to help my mate. Oh poor Trudy-wudy, he must be all tucked up in bed right now, sniffling away. If only I could bring him a nice cup of tea and a hot water bottle and kiss the Covid better.... anyway, I know he is in good hands and will be back on his feet before he knows it and ready to take on these nasty truckers once and for all.

In any event, in this chapter we've looked at some of the antivaxxer ringleaders and exposed them for the charlatans that they are. We then looked at how their ideas create vaccine hesitancy disease and examined in detail their brazen efforts to maintain the 'freedom' to murder us all. After considering all of these things, how can we sum up what these people are like? I would now like to offer some concluding thoughts...

Conclusion: The Unvaccinated are SELFISH!

If there were a study demonstrating that the unvaccinated are egotistical, selfish, blinkered, arrogant, suffering from near terminal cognitive dissonance and basically nothing better than armchair experts, then I, for one, wouldn't be at all surprised.

For the life of me I cannot understand whatever twisted mindset must be operating in someone who refuses to take the vaccine. Don't they know that they are putting everyone around them and indeed the whole of society AT RISK? That they might as well just walk around with a megaphone, announcing: 'I am a selfish shit because I don't care that the very air I breathe will probably kill you and all your loved ones'?

How can people be so SELFISH that they won't just do what we ALL know is CLEARLY best for them? It beggars belief.

Oh, bodily autonomy, blah, blah, blah, give me a break. What about MY bodily autonomy which will most likely suffer terminal consequences when it breathes in unvaccinated airborne droplets? Do these people never think of that possibility?! In fact, I would say that it is only a matter of time before the science shows that unvaccinated spittle is life-threatening even when it isn't infected with Covid. These people are harbingers of death, plain and simple.

Furthermore, vaccines PROTECT the lives of all those who take them. Why is it the unvaccinated can't 'get' this and are still making the decision to murder us all?

'Oh, but I had Covid, I have antibodies, why would I take the vaccine', some of them are happy to parrot. These people are tricky, I tell you, often making what appear to be totally sound, logical arguments but, in this case, all these people are really doing is revealing their seriously outdated beliefs in the value of the immune system. Furthermore, their priorities are all skewed. The obvious retort is: 'So you say you are safe from Covid and yet you don't want to be EVEN safer? You can never be too safe!'

And when they aren't killing all of us, the bastards are putting themselves at death's door from Covid (Hah! Serves them right!) and end up taking up invaluable intensive care beds. Is there no end to their egotism, I ask you? Here we all are, engaged in a society-wide and compassionate effort to save lives, and these loons have the audacity to end up near death and to take away resources from those who need medical care the most!

Indeed, these people hold unacceptable views and should not be tolerated. Most of us right-thinking people get this, as was demonstrated by the study from Aarhus University which showed that vaccinated people despise the unvaccinated (but bizarrely this wasn't reciprocated, God, typical doofuses that they are, truly unable to read the room!). For, the sad fact we have to acknowledge is that the anti-vaxxers are extremists and just like any other kind of terrorist, they need to undergo re-education. That's why I was personally over the moon when I read of the forward-thinking recommendations put forward by a psychology professor at the University of Bristol who said that those who refuse the jab should undergo deradicalization training.[12]

You see, the rest of us, we are the people who see the true need for social solidarity and cohesion. I don't know if you saw the beautiful German video made by some sheep farmers who put all their flocks together so that, from an aerial view, they formed the shape of a vaccine. The video was so poignant that it moved me to tears. Truly, the good eggs in this story are like those sheep, all being guided to take the vaccine for their own good and the benefit of all sheep everywhere.

And to take the vaccine is for our benefit, make no mistake about it! That's why the next chapter is so very important for we come now to the kinds of myths that the antivaxxers will spread about the vaccine itself...and so let us get down to the most important myth-busting of all! Onwards!

---

[12] I wish to put on the record, as Provost of the Termonfeckin Institute of Expertise, that I've generally been awe-struck by the developments emanating from Bristol University of late, a hitherto grossly underappreciated educational establishment. For example, they were recently the first university in the world to make clear that they will not tolerate discrimination towards any student or member of staff who self-identifies as a cat, i.e. those who are 'catgender'. No doubt such students will have separate pebble box toilets where they can at last answer nature's calls as nature herself intended. Indeed, Bristol's example has given me the intention to reflect upon how we too, in the T.I.E., can become more accommodating towards those of our students who are part of the CDLWQ+ community (that's the CatDogLynxWolfQuestioning + community in case you didn't know already, you bigot).

CHAPTER EIGHT: BUSTING ANTI-VAX MYTHS!

Right, we have now exposed the character of these antivaxxers, their deranged and weird attempts to fight for 'freedom' and the nefarious consequences of their actions. But what kind of things do they actually come out with about the vaccines themselves?

Now, all of the lies which follow in some way suggest that the Covid vaccines cause harm. Yes, I know, I know... the very same vaccines that have been tested rigorously and from every possible angle by top scientists and governments everywhere and which have been found to be safe and effective and, quite possibly for all I know, even to confer health benefits far beyond protection from Covid (well, it wouldn't surprise me anyway...in the same way that an apple a day is a damn good idea, no doubt a booster a year will have a very positive impact on your longevity).

Well, be prepared for a good ol' laugh when you see the kind of rubbish they will spew out about these life-saving vaccines! In particular, I will be focussing on four of their most pernicious ideas, namely that the Covid vaccines can: damage or edit our genetics; kill us (!); damage our hearts or cause a heart attack and, finally, that they can affect fertility.

So, let's now take on each of these in turn!

A Dangerous Gene Therapy?

I've already mentioned the extraordinarily innovative nature of the primary Covid vaccines used to date.... little messengers that go to our RNA genetic system and actually instruct it to make Covid's spike protein to which our body then creates an immune response...utterly genius altogether. But on the basis of this awesome technology, the antivaxxers dare to suggest that our genetics risk becoming damaged in some way.... well, lol, that's a bit of a leap, isn't it?

Well, you'd think so, but the sneaky creatures that they are like to suggest that scientific research actually shows this to be true!

For example, they like to point to a study by some supposed scientists at Stockholm University with a rather wordy and pretentious title ('SARS-CoV-2 Spike Impairs DNA Damage Repair and Inhibits V(D)J recombination in vitro'). In plain English, what this study was basically looking at was the effect that the spike protein, as created by the vaccines, has on DNA *in vitro* and what the researchers 'claim' to have found is that it actually inhibits DNA repair. Now, if you were conspiratorially minded, this would no doubt be interpreted as very concerning and as risking potentially serious consequences at some point or other down the line...why are we vaccinating our children with these things, yaddah yaddah, you know, anti-vax manna from heaven essentially.

Indeed, I honestly don't know what is going on with some scientists these days...do they not know that this is exactly the kind of study that the anti-vaxxers would just love to get their greasy hands on? I mean, just read the authors' conclusion that their findings 'underscore the potential side effects of full-length spike-based vaccines'. With answers like that, it should be clear to anyone with half a brain that these questions shouldn't even be asked in the first place.

But, thankfully, we have experts like myself to expose the serious flaws in research such as this!

First of all, it is from Sweden. Now, the Swedes used to be all cool and liberal but that changed forever with their 'no lockdown, no masks, let's murder our grannies' approach to the pandemic. Therefore, I would seriously doubt the credibility of any research emanating from such a place (and they also all love ABBA so case closed).

Secondly, their supposed findings were 'in vitro' rather than 'in vivo'. In other words, this study was done on genetic material outside a living human body. So therefore there is ZERO evidence to think that the same thing has now happened in the bodies of billions of people worldwide.

Indeed, it wouldn't surprise me if this so-called research team now wanted to conduct the same experiment in a living human body...Hah! As if you'd ever get the ethical approval to conduct *that* kind of study! The very notion strikes me as potentially extremely dangerous for human health. I mean, imagine if you did this experiment in just one person and you had the same findings, well, you'd have to stop the vaccine roll out immediately and then many millions would never receive the vaccine's health benefits! What a total travesty that would be. I, for one, am glad that potentially dangerous experiments like this Swedish one have remained purely in vitro rather than being carried out on large or wholescale groups of people which, in the circumstances, would be an utterly reckless thing to do. On that, I'm sure we can all agree.

Finally, even if these findings are true...what's the harm anyway? Lots of things damage DNA repair. Pesticides, chemicals, you name it, and sure most people don't give those things a second thought.

So I'd suggest that you don't worry your sweet head about it either.

I would, in fact, advise you similarly when it comes to another standard anti-vaxxer trope on the genetics front, namely that the Covid vaccines are not only damaging our genetics but actively changing or 'editing' them (honestly, it never ends with these people, does it?!)

Of course, they were all truly delighted with themselves and generally over the moon when a paper came out from Lund University (and yes, that's in Sweden too...can you see the pattern here?) which appeared to indicate that DNA alteration could well result from the Covid vaccines. Now, this paper has an awfully complicated title as well ('Intracellular Reverse Transcription of Pfizer BioNTech Covid-19 mRNA Vaccine BNT162b2 In Vitro in Human Liver Cell Line') but that's no real surprise, as it gives these kinds of people lofty intellectual airs, you see (or so they think anyway). In any event, what the paper claimed to find was that the RNA from the vaccines can 'reverse transcribe' (whatever the hell that means) to become DNA within the nucleus of the cell and in just a few hours post-vaccination at that. Basically, what I think they are trying to say is that the vaccines are capable of creating totally new DNA out of the RNA that is already in them, or something like that anyway, God only knows to be honest, I find it hard to enter into the mindsets of such people.

Ok, ok, so all along the anti-vaxxers have been going on and on about how the vaccines can edit your genetics and here, then, is a study which appears to show that they do just that...but hang on there, not so fast, please! For a start, the authors of this paper state quite clearly that they don't know if this new DNA remains permanently in the genome or if it just degrades and disappears. So yah, boo, suckers! It is by no means confirmed that our DNA is altered permanently by these vaccines *at all*! And, I have news for you, even if it was changed forever, what the hell is so wrong with having some new DNA anyway? These kinds of people make it out as if DNA is a bad thing! Frankly, it all sounds rather cool to me but then again, I'm neither a Luddite nor some dinosaur living in a bygone age... instead, I embrace science and all that it entails!

Ok, so much for the supposed effects of the vaccines on our genetics, what about all that 'death jab' nonsense? Well, the idea that the vaccine might kill you comes mainly from mentally ill misinterpretations of vaccine adverse event monitoring systems and so let us now turn our factchecking attention in that direction...

VAERS Reported Deaths and Adverse Events: A Lot of Hoo-Hah About Nothing

If I could have a free vaccination for every time the anti-vaxxers come out with their disgusting lies about the VAERS data, I'd be needled all over and full of health.

But what is VAERS, you might ask? It's the 'Vaccine Adverse Events Reporting System' and all those anti-vaxxers who claim to have died, or whatever they think happened to them, can use it to report their so-called adverse reactions. It's run by the US government (though really I think they should know better: you shouldn't encourage these kinds of people).

I don't normally like to present the anti-vax position in their own, twisted terms but I will in this case since it is evidently so ludicrous as to be laughable. Indeed, I actually found myself interviewing an antivaxxer (whom I found in God only knows which dark corner of the internet) just for this purpose so that you can see exactly the kind of fluff they come out with. Here is a transcript of the interview (which was in about March 2022):

'Oisín: So, you, give me your web of lies.

Tin Foil Hat Person: Thank you for the opportunity. Well, the thing is that VAERS shows there have been over 28,000 deaths reported for the Covid vaccines so far and a vaccine is normally pulled and investigated for safety concerns if it has had just 50 deaths associated with it and that is rather a smaller number than 28,000....

Oisín: Oh, just trying to show off your mathematical prowess, are you?!

Tin Foil Hat Person: No, not really. It's more that this suggests a concerning 'safety signal' you see and....

Oisín: Hang on a minute! It's us lot who want to save lives and keep everyone safe. How dare you appropriate our terminology?!

Tin Foil Hat Person: Well, then, why don't you care about helping those who are vaccine injured as well as those who are harmed by Covid? Why can't we do both? There have been almost three times the number of deaths associated with the Covid vaccines than from all other vaccines combined since records began...and over 150,000 Covid vaccine related hospitalisations....

Oisín: Oh, trust you to be counting! But why should we take these reports seriously? I mean anyone can do one.... You've probably made up at least half of the fatality reports for all I know!

Tin Foil Hat Person: Well, no.... most of them are completed by physicians and it is actually a crime to submit a false report.

Oisín: A crime to submit a false report?! I'd make it a crime to submit a report in the first place!

Tin Foil Hat Person: We should also remember the study that suggested only around 1% of vaccine adverse events are even reported in the first place...

Oisín: Hah! I bet that was a 'study' by one of those anti-vax, 'save the children' or whatever groups....

Tin Foil Hat Person: No, it was by the Centers for Disease Control and Prevention, the CDC...

Oisín: Yeah right it was! Liar, liar, pants on fire!

Tin Foil Hat Person: Why would I lie about that? Similarly, the German government just released data suggesting that 1 in every 5000 Covid vaccine doses results in a 'serious' adverse event, where that means someone's reaction is severe enough to result in hospitalisation. Now, that surely should be of concern to us all and...

Oisín: Oh so the German government goes all anti-vax and we're supposed to take them seriously, is it? Ach, I'm so bored of you now... this interview is over!'

My God, could you have greater evidence that we are dealing with paranoid nutters than this total dribble? Indeed, speaking of Tin Foil Hats, I think that what we need is for the next edition of the DSM (Diagnostic and Statistical Manual of Mental Disorders) to have TFHPD (Tin Foil Hat Personality Disorder) along with clear guidance that treatment is impossible and imprisonment the only feasible option.

I mean just look at what my loopy interviewee was coming out with... isn't it obvious that these people are totally unaware of a basic principle within the scientific method, namely that *correlation is not causation*. Just because Auntie Mary has her vaccine, starts frothing at the mouth and then drops dead half an hour later, does not mean that the vaccine *caused* her death. It could have been that the chocolate biscuits she ate that morning were a bit off or that the toothpaste she used during her morning ablutions had been poisoned by her neighbour who was fed up of hearing her natter away on the phone through those paper-thin walls. You just don't know what's really going on and you should never jump to conclusions too quickly. Just because it looks like a duck, quacks like a duck and waddles like one does not mean that it is not an elephant.

The second thing is that, whatever this nutter was saying, I very much doubt that it is a crime to falsify VAERS data. My own impression is that the anti-vaxxers are just conspiratorially beavering away in their underwear, submitting false reports every day. They read obituaries in the paper and they make up a report. They see someone who has died on social media and they make up a report. For all I know, they probably go out and murder someone, poking them all over with syringes, and then make up a report. I think these scenarios are far more likely than the idea that these reports are genuine. And sure why would a physician ever report such adverse events: after all, physicians who do so are often threatened with losing their jobs and being struck off, and why would any doctor want *that* to happen? It just doesn't make any sense!

So, I hope I have shown you how to counteract the kind of irrational statements that Tin Foil Hat people will come out with about VAERS and similar reporting systems. These people just do not have the critical thinking capacities to interpret the data within such systems correctly. To think that a case in which someone's death is attributed to a Covid vaccine actually means that that person died of that vaccine is really lowest common denominator thinking. But sadly that, folks, is just the kind of nonsense that we must contend with.

But how exactly are the vaccines supposed to be killing us all? Well, in practically every way under the sun, if you were to believe the antivaxxers. But I can't possibly focus on every supposed cause of death as we'd be here until the cows come home to roost, as the old saying goes. Instead, therefore, I will home in on one of the main ones, a particularly egregious myth of which most people have no doubt heard, namely that the vaccine is possibly not so great for the old ticker...

No, the Vaccines Will Not Give You a Heart Attack!

Now the same 'correlation is not causation' point also applies to the amount of cardiac events registered in VAERS. You can be quite sure that the 15,751 supposed heart attacks and 50,176 cases of myocarditis/pericarditis (as of July 2022) are nothing at all to do with the vaccine, whatever the anti-vaxxers might tell you.

But, again, the slippery customers that that they are like to suggest that the science is on their side with this one too! And if there is one paper that they champion for this particular cause it is one by a certain cardiologist by the name of Dr. Steven Gundry with the title 'Observational Findings of PULS Cardiac Test Findings for Inflammatory Markers in Patients Receiving mRNA Vaccines'.

So what this Dr. Gundry did was to look at various blood markers (IL-16, soluble Fas and Hepatocyte Growth Factor - yeah, I know, I'd never heard of them either!) in patients before and after Covid vaccination. And sure why would he do such a thing, you might ask? Well, according to him (but who'd believe him?), these are signs of endothelial and vascular damage and can be used to predict heart attack risk over a five-year period. He compared the levels of these markers before two doses of the vaccine and again at two weeks and three months after. What he found is that the heart attack risk for a five-year period rose from an average of 11% to 25% among the 566 patients studied.

Now much hoo-hah has been made of this among the anti-vaxxers with many of them claiming that this is a 'shocking increase' and 'extremely concerning' and the like. Well, that's all very well, but the actual fact is that a 25% risk of having a heart attack is actually the same as a 75% risk of NOT having one. And, if you ask me, those are still pretty good odds. Clearly the real finding of this study should be that 'After Covid vaccination, people are STILL more likely than not to remain free of a heart attack over a five-year period'.

All that said, I very much doubt Mr. Gundry's findings here. Sure, if what he is saying is true, then it would be very dangerous for us all to have annual boosters for the rest of our lives, wouldn't it? And that can't be the case because the heads of all the vaccine companies say these vaccines are safe and effective and I've no reason to doubt them. I mean, with a 25% risk of a heart attack after two doses, what would it be after your seventh shot? 65%? The very idea is outlandish. Sure, we'd all be dropping like flies. I don't believe it. For all I know, Mr. Gundry is making these blood markers up.

And, furthermore, even if someone was concerned about an increased heart attack risk post vaccine well, frankly, they shouldn't be. After all, the same companies who make the vaccines also make drugs for cardiovascular problems, so you'd be well looked after.

And while we are talking about heart attacks, we might as well deal with the heart inflammation myth as well and particularly the idea that young men are at a significant risk of developing this post vaccination.

And, to be honest, this is a really typical one. All over the internet you'll find Republican mums talking about their little darling Bubba or Linus and how, there they were playing high school baseball one minute, and the next unable even to get up the stairs and confined to their beds.

But the expert view is that these are nothing other than 'mild' cases of heart inflammation and easily treated, the medical equivalent of having to take an aspirin or a strepsil. Indeed, I don't know about you but, ever since I was young, I've always heard of neighbours and family coming down with a spot of the old mild heart inflammation.... sure there is nothing unusual about it. I distinctly remember my mother saying: 'Oh did you hear, Oisín, Auntie Carmel has some mild heart inflammation at the moment. She'll be popping along to the doctor today to discuss it along with her mild appendicitis.'

So when you come across studies like the one from Hong Kong which suggest that one in 2700 teenage boys end up with heart inflammation after their 2nd dose,[13] something which the antivaxxers decry as an 'alarming risk', it is important to remember that what these young men are facing is totally fixable. And, in other ways, this heart inflammation arguably carries benefits as you could firstly say that these lads are learning the importance of 'taking one for the team', a valuable life lesson that will keep them in good stead for the rest of their lives, however long (or short) a time that might be. And, secondly, sure the hearts of these young men are inflamed enough with the lusts of youth as it is and that's much more likely to run them into trouble,

---

[13] 'Epidemiology of Acute Myocarditis/Pericarditis in Hong Kong Adolescents Following Comirnaty Vaccination'

if you ask me. So it's good for them to have a bit of a break from all that carry on, if it is not too old-fashioned of me to say so.

Anyway, there you have it, my dear readers, no, the vaccines will not make your heart explode, on that we can all be 100% certain. But there is one last and most insidious myth to deal with, one of the sickest if you ask me, namely that the vaccine can affect the fertility of the fairer sex... honestly, can you believe the kind of thing these people are capable of coming out with? No, me neither, and that's even with me after writing this book about them! In any event, let's now deal with this particularly egregious lie....

No, the Vaccines Will Not Affect Your Fertility! (if anything they'll just make superhero babies)

Now, this particular myth stems from supposedly internal research documents of a leading vaccine manufacturer which showed that the spike protein created post vaccination travelled all throughout the body, concentrating particularly in the ovaries. Now, even if this were true, I can't for the life of me see how it could be a problem. I mean all it would mean is that babies are protected from the get-go and probably won't EVER need to be vaccinated themselves. So, as far as I'm concerned, this would simply be further proof of the amazing technology that underpins these vaccinations.... two for the price of one as it were.

I believe that the really sad news in all of this is that the same documents showed that the spike protein did not end up in the testes *at all*. This would appear to be a missed opportunity: imagine if your sperm could be a weapon in the fight against Covid! Spike infused sperm meets spike infused ovaries which equals babies everywhere born as little Covid-fighting machines and absolutely glowing with health.

For these reasons, I was very upset to read recently that the European Medical Agency has decided to investigate increasing reports of menstrual irregularities post vaccination. They really shouldn't kowtow to the antivaxxers in this way.... after all, it'll just be more grist for their elbows, as the old proverb goes. Perhaps I should send my fellow experts in the EMA a copy of my book so as to reassure them on this point.

So much for the misinformation about fertility. In general, when I think back over this entire chapter, it seems obvious enough to me that anyone who thinks these vaccines are not safe is truly living in doolally land. The fact is that even the youngest among us faces ZERO risks from them. And that's why government regulators, such as those in Australia, are increasingly decreeing that children as young as 12 need not obtain parental consent in order to be vaccinated. This is spot on although, to my mind, that age should be lowered even further. For is it not the case that a new-born babe, just as it reaches naturally for its mother's breast, would also reach for the syringe?

Well, I think so anyway. Indeed, it seems clear enough to me that the real issue here is not with the vaccines but with the mental health of the antivaxxers. Is not the very definition of a hypochondriac someone who is still worried incessantly about their health even when their doctor has told them there is nothing wrong with them and that they are just a bit stressed out and need to take it easy? And therefore what, I ask you, is the difference between such anxious types and what we witness with the so-called 'vaccine-injured'?

Nothing that I can think of. I mean, I don't know about you, but if I go to my doctor thinking I have the symptoms of heart inflammation or paralysis, there is nothing like this kind of reassurance to help calm me right down and feel a whole lot better. But with this lot, they honestly have the audacity to ignore their doctors' advice and instead take to the internet and broadcast their woes to the world. This is the attention seeking behaviour typical of malingerers everywhere.

I can say this because the science shows that the only side-effect these vaccines can ever really give you is a bit of tenderness at the injection site. ANYTHING beyond that is simply a touch of hypochondriasis.

For example:

If someone is complaining of uncontrollable shaking and seizures, well, we all know that anxiety gives people the shakes.

If someone is complaining of heart issues, well, we all know that anxiety leads the heart to flutter and race.

If someone is complaining of bladder incontinence, well, we all know that anxiety leads us to wee and widdle more often.

If someone is complaining of being paralysed, well, we all know that anxiety can 'paralyse' us with fear.

If someone is complaining that they can't walk, well, we all know that anxiety can make our legs like jelly.

As for those who claim to have died from the vaccine, well, we all know that you can die of fright.

And sometimes if people are really anxious, such as these anti-vax hypochondriacs, they can have all of these symptoms at once. Typical.

Well, we have now exposed the most common anti-vax myths about the supposed dangers of the Covid vaccines. You'll know well what to say to one of the loonies the next time they are talking about the 'risks' of using such a novel technology, the VAERS data or the heart attacks people are supposedly having left, right and centre (like all the footballers, apparently, although they are obviously keeling over just because they are super unfit after spending lockdown loafing on the couch).

Where to next? Well, given what these antivaxxers are like and their proneness to health anxiety, it'll probably come as no surprise to you that they are also real pill poppers. Well, not proper medications of course.... I'm talking about whatever latest fad supplement has hit their health food store. No doubt they think these things 'cleanse their auras' or what have you... but they bring this same propensity towards swallowing any old thing to dealing with Covid infections. You see, at the end of the day and despite everything, they are also frightened of Covid, just like everyone else, and so they have developed their very own Covid quack medicine cabinet, as we shall now see....

CHAPTER NINE: QUACK COVID CURES

One of the aspects that leaves me most speechless about Covid-deniers is that, despite the fact they are happy to cause countless deaths through their pandering of vaccine misinformation, they nevertheless also claim they want to save lives from Covid just like we do (I know...you couldn't make it up!) But what kinds of Covid 'cures' do they come up with? Well, as you'll not be surprised to learn, it's the usual unthinking cognitive diarrhoea that we've all come to expect from this lot and, to (back)kick this chapter off, we will first consider something which (clip clop) really epitomizes (cloppity clop) their whole approach (neiiiigh! neeiiigh!), namely their certain fondness for yup, you guessed it....Ivermectin!

Ivermectin (because Covid obviously gives you worms! And turns you into a horse.... Jeez, the intellectual level of these people!)

The other day, my neighbour Máire gave me a buzz. 'Oisín', she whispered down the phone, 'Did you hear that Ivermectin has come to Termonfeckin? That old sheep farmer Séamus is after stocking up on it, says it will help him fight off the China virus. Can you do anything about this before other people get the same idea?' 'Oh I absolutely can and will, Máire, thank you for bringing this to my attention!'

Three days later, I published an exposé of Séamus as the lead article in *The Termonfeckin Tribune*:

> 'Termonfeckin Tragedy Averted as Totally Thick Local Sheep Farmer Exposed!
>
> Séamus O'Shaughnahoy, a sheep farmer for four decades, has been found by the Gardaí to be in possession of 12 packets of Ivermectin, a horse deworming drug that conspiracy theorists claim works as a treatment for Covid.

Remanded into custody, Séamus is currently awaiting trial under charges of conspiratorial thinking and gross stupidity. *The Termonfeckin Tribune* was given access to Séamus in his cell.

TT: Séamus, are you not a total thicko for falling for all that horseshit?

Séamus: It isn't horseshit, nor is it a horse dewormer. It's a Nobel Prize winning drug that has been successfully repurposed for Covid and....

TT: The only prize anyone will be getting in this situation is you for being so utterly empty between the ears, The Termonfeckin Thicko Award!

Séamus: No, it is true what I say! For example, it's been shown to have anti-Covid properties and has been used to great effect in Mexico and India, among other places.....

TT: Oh God, listen to you! You're talking like you are an expert! It's a horse pill, you eejit. And it'll be no good to you now, not even for your sheep, or did you expect it to work for them too?'

Having done my citizen's duty in this way, I was mighty proud of myself and felt that I had helped avert a serious crisis in my own hometown. However, I admit that a little voice was nagging at me at the back of my brain, wondering what on earth he was talking about with Mexico and India. Could they really be so mentally soft over there that they would also fall for this kind of tin foil hat thinking? So I decided to have a look into what the story was and, lo and behold, what I found was more fodder[14] for all my myth-busting efforts and so, here we go, folks...

First up, Mexico. It would seem that what Séamus was talking about was a study by The Mexican Institute of Social Security, headed up by one Cesar Raul Gonzalez-Bonilla. Now, what this Gonzalez-Vanilla came up with was the idea of seeing what happens if you send a

---

[14] Hilarious, Oisín! (Ed.)

'home treatment kit' out to people in Mexico City, a pack that included a course of Ivermectin, and compared the outcomes in this group to those who did not receive a treatment pack. In all, 28,048 people who ended up with a confirmed Covid diagnosis were tracked. The results showed that 11.71% of the non-Ivermectin group were hospitalised in contrast to 6.14% of the Ivermectin group.

Ok, ok, so I know what you are thinking.... were these people or horses who got Covid? That's the first objection that came to my mind also and, sadly, from looking through the study with a fine honeycomb, I couldn't find the answer to this question anywhere.

But can we really say that these results indicate that Ivermectin is all that helpful? Maybe the extra five percentage points in the Ivermectin group who didn't end up being hospitalised actually ended up neighing and galloping around their neighbourhood all the way to the nearest psychiatric institution? I mean, I don't know but it seems possible to me, anyway. Until these kinds of questions are clarified, I personally wouldn't give too much weight to this study by Dr. Gonzalez-Gorilla.

Next up India, and the state of Uttar Pradesh, which, from my internet perusals I can safely say has achieved some kind of cult-like status among those of us who lack critical faculties.

So what's the rub here, then? Basically, the Uttar Pradeshian State Health Department took part in a most dangerous experiment early on in the pandemic giving Ivermectin preventatively to all health care workers. According to State Surveillance Officer, Mr. Agrawal: "It was observed that none of them developed Covid-19 despite being in daily contact with patients who had tested positive for the virus." Well, all I can say, is they clearly got lucky, but on the basis of this luck, they then had the audacity to suggest a State-wide Ivermectin program! Close contacts, healthcare workers, Covid cases, everyone was to take Ivermectin in what the State termed a 'prophylactic and therapeutic' program.

For me, their very wording here shows up the seriously low level of competence among these health officials. It's only a short leap really, if you are suffering from being intellectually challenged, to think that

Ivermectin will not only save you from Covid but also allow you to have it off with impunity! Honestly, I ask you: where on earth are the morals in all of this?! I might be old fashioned but I don't think that it is the place for ANY government health department to be suggesting that people should be going at it like rabbits (or, indeed, like horses, for that matter).

And it is in this context that we should interpret the 'results' of this program. Mr. Agrawal goes on to say: 'Despite being the state with the largest population base and a high population density, we have maintained a relatively low positivity rate and cases per million of population.' So yes, as of writing this in April 2022, Uttar Pradesh, with a population of 204 million, has had 23,494 deaths while another Indian state called Kerala, with a population of just 35 million, has had substantially more at 67,772 deaths and that would indeed very much seem to be a feather-duster in Uttar Pradesh's earlobe, wouldn't it? But since we now know that it was all just a state-sponsored excuse to have an orgy, we can deduce that the real reason for the low case numbers is that everyone was just staying inside and having a good time, thereby limiting the spread of the virus very effectively.

So as for Uttar Pradesh, I can say with utmost confidence is that it is all Utt-ar Rubb-esh![15]

Anyway, I now know what I will be saying to Séamus the next time I see him (when his term is up, that is). Indeed, he shouldn't go researching things like approaches to pandemic management in other countries, it'll only mislead him and give him a headache when he needs to preserve as much of his rather limited mental nous as possible for his sheep shearing.

Finally, there was another study recently which settled the Ivermectin question once and for all. Looking at a massive database of patient outcomes across the US, it compared mortality rates among those treated with Ivermectin with those treated with the super-duper US government approved Remdesivir. The title of the paper, indeed, tells

---

[15] Oh you are on fire today, Oisín! (Ed.)

you all you need to know: 'Treatment with Ivermectin Is Associated with *Increased* Mortality in Covid-19 Patients: Analysis of a National Federated Database'. So there, losers![16]

Anyway, Ivermectin represents the pinnacle of anti-vaxxer thinking on how to cure Covid. And you'd think it couldn't get any worse but really the rest of their offerings in this regard are even less impressive. I mean at least Ivermectin actually *is* a medication (even if one just for horses). The rest of what they come up with truly does belong to the realm of alternative health...don't they realise just how serious Covid is? Well, clearly not, and so on we go to...

The Vitamin D Debacle!

Of course, if you think about it, it is no surprise that the Covid-deniers latched onto an animal pill as their top Covid treatment option.... they aren't exactly the brightest sparks, after all. And their next 'solution' for Covid really just continues with that theme. You see, Covid-deniers are very fond of their natural health cures and supplements of all sorts (it's probably only a matter of time before they come up with a Colonic Covid Cleanse or some such). Therefore, it was really no surprise to me when they started to go on and on about one of the dirt-cheapest supplements available, namely Vitamin D.

Oh, Vitamin D will boost your immune system and isn't that a good idea when the immune system could do with being boosted, etc, etc, etc. Listen here, you quacks, the only booster your immune system needs is your 17th jab, so don't you give me that crap. And it is total crap as I will demonstrate right now.

Indeed, I went to do an online search for links between Vitamin D and Covid outcomes and there are literally hundreds of papers at this point suggesting that the lower your Vitamin D the more likely you are to die from Covid.... but, to be honest, I don't have the patience for reading through THAT much research by far-right scientists so

---

[16] Um, Oisín, it was actually 'decreased' mortality. Should we really leave this one in? (Ed.)

here's just one example. There was a German study[17] which looked at the Vitamin D status of Covid patients among those who ended up in intensive care and among those who ended up dying from the illness and what these so-called researchers claim is that those with low Vitamin D levels were 15 times more likely to require an ICU bed and six times more like likely to die from Covid. Well, I suppose if you were a bit mentally challenged, those kinds of conclusions would lead you to think that governments should mandate everyone to take a spot of Vitamin D every day…. cheap, easy, safe, life-saving, blah blah blah, you know the kind of carry on by this stage.

Well, not so fast! Let's just ponder the implications of this suggestion carefully, shall we? Would we honestly be comfortable with governments mandating that everyone should take a supplement? What right in God's name does the State have to control what goes into our bodies?! Sure, we don't even know what else is in these supplements! The whole thing could result in a total society-wide health disaster, Vitamin D for Vitamin Death, the greatest public health debacle in history. And then, even if you did implement it, how on earth would you enforce it…. *oh you can only go in here if you are Vitamin D sufficient, show me your Vitamin D papers*. The whole thing would be total madness and that's why it doesn't matter what these studies show, the implications of them are totally and utterly unworkable.

Furthermore, whatever all these studies might say, I'm really not sure that Vitamin D is that big a deal when it comes to Covid.

In fact, I have another study up my sleeve, from my own dear country in this case, which I believe casts serious doubt on the idea that vitamin D is all that helpful for the Covid (the study was called 'Vitamin D and Inflammation: Potential Implications for Severity of Covid-19'). Now what this study did was to compare Covid outcomes by Vitamin D status between countries in the north of Europe, which receive little sunlight, and sun-filled countries in the south of Europe (interestingly the Scandinavian countries actually had higher vitamin D levels despite receiving less sunlight, probably because their

---

[17] As reported in *Irish Examiner*, Feb 19th, 2021 ('Vitamin D: Can the sunshine vitamin help to put Covid in the shade?')

governments fortify their food supply with it, or so I have been told). So, did the researchers find that the vitamin-D sufficient northern countries had lower death rates than the vitamin D deficient southern countries? Why, yes, they did but don't go thinking that that means the Covid deniers are right just yet...wait till you've read my own expert take on this study. But, first of all, this is what the researchers said:

> 'Counter-intuitively, lower latitude and typically 'sunny' countries such as Spain and Italy (particularly Northern Italy), had low mean concentrations of 25(OH)D and high rates of vitamin D deficiency. These countries have also been experiencing the highest infection and death rates in Europe. The northern latitude countries (Norway, Finland, Sweden) which receive less UVB sunlight than Southern Europe, actually had much higher mean 25(OH)D concentrations, low levels of deficiency and...lower infection and death rates.'

Ok, so, yes, as I mentioned this study would appear to show that a higher Vitamin D status does indeed correlate with lower death rates from Covid. But is it really the higher vitamin D levels that are the main reason, the 'active ingredient' as it were, that is responsible for the lower Covid death rates in Scandinavian countries? I'm not so sure we can infer that from this study AT ALL. Indeed, there is a whole other potential factor in this case which the study's authors refer to but which, bizarrely in my view, they do not highlight as completely essential for their analysis. And what is that factor? Well, in my expert opinion, what this study *really* highlights is the hitherto grossly underappreciated fact that countries which receive *less* sunlight like Norway or Finland also experience *lower* Covid death rates. In other words, I don't think vitamin D status has anything to do with the matter at all. Rather, it's the amount of sunlight you get that determines your Covid outcome and the less you get the less likely you are to die from it! What this study is really showing, therefore, is yet another way that sunlight can kill you.... first it was with cancer and now it is with Covid!

And so isn't this therefore nothing other than a TOTAL vindication of government strategies everywhere of ordering people to stay indoors and away from the sun as much as possible? This strategy becomes doubly genius when you add in mask mandates as then, on the odd occasion when you do have to go outside, you are at least blocking even more of the sun's rays from touching your face and therefore making you even healthier and protecting you even more from Covid.

So, as a last word on this Vitamin D nonsense, folks, please just follow the science and STAY inside.

Well, there is time for us to consider one last quack Covid cure and this only goes to expose the health charlatans that are amongst us even more. Indeed, what could be less scientific than the idea that what we eat could possibly have any influence on such an horrific disease as Covid? Not much, I would say, and so let us now consider this completely pernicious idea which really and clearly falls into the category of being a total joke...

No One Will Take Away My Right to Eat Ice-Cream!

There are few things in life that make me happier than ice-cream and preferably one laden with cookie dough and bits of chocolate cake. In fact, I'm slurping away on some right now and really delicious it is too.

Is there anything more innocent than such pleasures? And yet some of the conspiracy theorists would have us believe that scoffing all these sugar-laden mountains of joy is more likely to make us die from Covid! Good God, talk about fear-mongers and prophets of doom....as if such sources of happiness could ever do such a thing! It really says a lot about the mentality of these people that, when all of us are engaging in a valiant act of self-preservation by staying indoors and baking chocolate cake, they tell us that we'd be better off eating healthily and getting outside for some fresh air. Talk about people who have their life priorities totally arseways.

So let's now obliterate this nutty idea that lifestyle interventions should in any way be part of a government's health response to Covid once and for all. As always, let's start off with debunking the kind of 'evidence' the conspiracy theorists make use of.

One example is a study from Tulane University published in *Diabetes Care*. It found that those with 'metabolic syndrome', a term I'd never heard of in my life but which apparently is typified by high blood pressure, high blood sugar / diabetes, obesity, high triglycerides and low HDL cholesterol, were 3.4 times more likely to die from Covid and five times more likely to enter ICU and this is something to do with how these conditions make Covid more likely to get into something called the ACE-2 receptor or some such, also something I'd never heard of to be honest.

In any event, I don't even know where to begin with this study. For a start, it is evidently 'sizeist' in its picking on our chubbier friends. I mean, it takes a long time for fat people to accept and love their considerably large selves and then along come studies like this that try to suggest that being overweight is unhealthy. What shameless stereotyping! Why don't the authors go and pick on someone their own size?

Secondly, so what would these researchers suggest, that people should be encouraged to eat less chocolate, cake and crisps and eat more fish and veg? The very idea that a government could so interfere in people's lives strikes me as both utterly appalling and as an encroachment on the most fundamental civil liberties! I'm no philosopher but if there is one thing in life for sure that I know leads to happiness it is ice cream and no government will EVER take away my freedom to eat as much of it as I want. And if they did, you could be sure I'd be on the streets protesting along with my fellow Ben & Jerries loving friends. Well, we wouldn't want to exert ourselves too much but, for sure, we'd be at the very least occupying the city centre with our cars, honking away and letting our general displeasure be known loud and clear.

Anyway, it takes ages to reverse health conditions and that's if it's possible at all, frankly. Though, again, these nutters will point to all sorts of things. One is a study from Italy of all places, 'Middle and Long-Term Impact of Very Low Carbohydrate Ketogenic Diet on Cardiometabolic Factors' which followed the effect of a low carb diet on 377 patients over one year and their body weight, blood sugar, blood pressure, lipid levels and glucose metabolism. What the study claimed to find was a 'significant improvement' in all these areas.

Look, we live in an era of fake news and the very idea of such a study emanating from the land of pizza and pasta strikes me as downright odd. But even if it is true, and even if these health parameters did all improve, what about other aspects of these patients' health which were not studied, eh? For example, what if the real finding from this study should have been: '377 patients forced to undergo diet of near starvation lose weight but also now suffer from major depressive disorder and suicidal ideation due to absence of regular tiramisu and panettone.' Now, that would put a different spin on things, wouldn't it?

And even when you look at studies like those of Phinney & Volek which claimed to show that a low carb ketogenic diet resulted in 147 out of 262 diabetes patients reversing their condition after just 10 weeks, really that's still an AWFUL long time to be sticking at something to get any results. Especially when it only takes a couple of hours to pop down for your 8th booster and then you are all set for another couple of months. In fact, if I were to acquiesce a little, I would suggest that the calories burned from the effort of getting to your local vaccination centre are certain to give a health boost to those who are overweight. And so we should do whatever is necessary to encourage these people to make the trip to get their vaccination... I thought the policy in some US states of offering people free candy and cake in return for the jab was the perfect incentive, in fact. There you go, Covid-deniers, that is a clear situation where eating cake is GOOD for your health! And a perfect example of a kindly government health policy, not just more carrot than stick, but, one could say, more cookie than stick.

And so, if you don't mind, I've got this tub of deliciousness to finish.

Chapter Ten: The Great Reset (or 'The Much Needed Plan to Save Humanity from Itself')

Well, my dear reader, we have almost come to the end of this book. It's been quite a ride, hasn't it? I've certainly had a lot of fun writing it. Indeed, I have taken the greatest of pleasure in tearing apart all of the most prominent anti-vax myths and I hope that you have taken equal pleasure in seeing their shaky premises being exposed once and for all.

But we aren't quite finished....no, indeed, you wouldn't be able to peruse a Covid conspiracy theorist's Twitter feed for long without finding mention of the supposedly nefarious dealings of the World Economic Forum (WEF) and their 'Great Reset' agenda. In this case, the idea seems to be that the WEF has been using its extensive network of current and former members, many of whom are now Presidents and Prime Ministers, to usher in a truly dystopian future in which all citizens have digital ID 'passports', social credit ratings and their lives micromanaged by global elites. Talk about Covid-19 becoming Covid-1984!

Sounds craaazy, right?

So, let's get down to our final round of myth-busting!

Klaus Schwab: A Wise Swami for Our Times

When it comes to the so-called 'Great Reset', most of the anti-vaxxer ire is directed towards the head of the WEF, Klaus Schwab. For years, they claim, this man has been influencing world leaders with his 'Fourth Industrial Revolution' ideas through the Forum's annual meetings at Davos and his 'WEF Young Leaders' program. Few anti-vaxxers won't have seen the clip of him in which he speaks of his pride in the success of this program and of its many graduates who have gone on to become Prime Ministers around the world (among them many of my poster pin-up favourites including my mate Justine Trudy, Jacinda Ardern, Angela Merkel and Emmanuel Macron, all of whom played an absolute Covid blinder it must be said). In this video, he speaks proudly of 'penetrating the cabinets', which to the conspiratorially minded would seem to be referring to some kind of ideological infiltration but surely, if anything, this is just a case of some innocent locker-room banter about the man's substantial virility? Well, all I can say is that he clearly looks after himself and must be quite the catch and if I were (a female leader, of course) in charge of a country and found myself the merrier after a few glasses of bubbly on the last night of the Davos Forum, well, I don't think anyone could stop me....

Anyway, as for the idea of a 'Great Reset' agenda, have you ever heard of anything more cooky or downright nutty? Nah, me neither! I mean the very term strikes me as completely bizarre. That's why when I saw one day on Amazon a book entitled 'Covid-19: The Great Reset', I thought it had to be by some prominent conspiracy theorist or other. But then I saw that the author was none other than Klaus himself and I was somewhat brought up short. Perhaps the great man was having a laugh at the antivaxxers, I thought to myself, appropriating their terminology and so forth in some kind of humorous satire exposing their lies but no, I went to read the book, and what I found was a hefty treatise on how the world does indeed need to be reshaped as a result of the Covid pandemic as well as many ideas regarding how this can be achieved.

Many questions swirled through my mind. Why did Klaus choose to use the exact same 'Great Reset' phrase that the antivaxxers had (surely erroneously) claimed was first coined by him? And why was he putting forth an agenda for a whole new kind of world when that is also exactly what the antivaxxers had accused him of doing? Was there something, I pondered long and hard to myself, *else* going on?

But no, I then concluded with absolute certainty, that simply can't be the case! After all, people such as Klaus and myself are the good guys! And sure, when you think about it, isn't it obvious that the world needs to be reshaped? And so I returned to read the entirety of Klaus' book with a more open mind and didn't I then fall in love with the lofty ideas therein and the doubting Thomas within me got put right at the back of the class where he belongs. There is no dark conspiracy here, oh no, that I can say for sure. Rather, this is the case of an enlightened being, a Buddha or a Jesus type, positing a roadmap which we can all follow to better ourselves. This is truly the blueprint for a beautiful world, one where we will all have sustainability coming out of our ears and will be full of the highest levels of wellbeing imaginable. Of course, the anti-vaxxers don't want that kind of thing, do they? The miserable sods that they are. They are happy for the climate to be burning up, for life to be all about greed and so on and couldn't envision a better world even if they tried.

Look, it is true that Klaus' book is full of extremely intelligent language and, to be honest, I would struggle to convey the majestic nature of his ideas adequately. But it isn't necessary to do so as the WEF has been publishing short videos for the last few years with bitesize take-home points so that the rest of us can understand just how The Great Reset will play out. And one of these videos in particular had the anti-vaxxers up in arms and, as you will see, for no good reason.

Now this video, describing the world in 2030, included the phrase: "By 2030, you will own nothing and you will be happy." And, typically enough, this perfectly benign command has been used as 'evidence' by those of a cooky disposition that the Great Reset is ushering in a truly dystopian future....as if there is anything wrong with wanting us all to be happy! Oooh, so evil and sinister....NOT! And more than that, I think if someone comes along and suggests there is a way for us all to be happy, we should take stock and listen very carefully, don't you? After all, Klaus and his mates strike me as the happiest of campers and, if they know the secret to a good life, I'd like to be let in on it... indeed, I would imagine that it isn't just any old happiness that they are talking about, but rather some kind of deep, long-lasting and complete happiness....

Let's just pause for a moment. Doesn't the idea of being 'completely happy' not sound pretty good? I mean who wouldn't want to be completely happy? As far as I'm concerned, if Klaus and his gang have worked out a way to keep our serotonin levels permanently ramped up, then I'm all for it. Dystopia, my ass!

And how is this 'complete happiness' to be achieved? Well, by owning nothing, of course. And how does that work, exactly? Well, this is where we have to bow down before the mighty philosophical principles that underlie this beautiful vision for our collective future. We all know that the pursuit of material goods, a.k.a. The American Dream, does not lead to happiness. We can therefore conclude that: having lots of things would make someone very unhappy, having quite a lot of things would make someone moderately unhappy, having a few things would make someone a little unhappy and having nothing at all would make someone completely delirious with joy. 'Take that Aristotle' is all I can say.

And so instead of owning anything, we will rent out everything we need as 'services' from certain major, multinational businesses. Now, the owners of those businesses will of course 'own' these things and thereby be the only people anywhere on the planet who still own anything. But given that this means that they will essentially own everything in the world, and because *not* owning things is what makes people very happy, we can only assume that these people will be *extremely* unhappy (as in 'down in the dumps, please kill me now' unhappy). What this really shows us is just how altruistic and unselfish these people are and how we should be grateful to them for sacrificing their own happiness so that we, instead, can all be truly joyful and generally in a state of near constant euphoria.

Is this not all this the height of philosophical insight into the human condition? Plato, the Stoics, Confucius, Lao-Tzu…. they all grappled with the question of human happiness and came up with their own answers, all of which pale in comparison, I'm sure you will agree, with the conception offered us by Klaus. And so, I, for one, will always be grateful to him for spotting the silver lining in the Covid cloud. But then, we should hardly be surprised. Did you know he published his tome on Covid-19 only four months into the pandemic? That he should have spotted the potential Covid offered us for wholescale change on a worldwide scale in such a short period of time…. is this not the greatest testimony to the man's industriousness and generally luminary nature?

Indeed, I really don't know how anyone, let alone the anti-vaxxers, could have a problem with Klaus Schwab. That lot might compare his looks to those of a Bond villain but, to me, the man has kindly, benevolent features, twinkling eyes and a real desire to bestow gifts upon humanity. In fact, my own affectionate nickname for him is 'Santa Klaus', so much do I feel that he has our best interests at heart. I could certainly see him in a Santa suit, children at his feet and on his knee, and him asking them which booster they want for Christmas.

Well so much for the idea of a dystopian 'Great Reset'. I hope you can now see how this is only something to be embraced. But, in addition to the beautiful vision contained within Klaus' ideas, I myself believe that I can also add a few more ideas into the mix for how to make the next period of humanity a truly golden one. Do you remember back at the beginning that I mentioned I felt the need to make a great ejaculation? Well, it is to this, at last, that I now come...

The Great Termonfeckin Ejaculation

Didn't I just read the other day that none other than Mr. Bill Gates himself is after writing a book about how to deal with the next pandemic and that the WHO is asking all countries to sign up to a pandemic treaty? Similarly, I hope that the following five points which we, at the Termonfeckin Institute of Expertise, have devised at great length can feed into the sterling work put forward by the luminaries in the WHO, Mr. Gates and, of course, dear Klaus and all his friends at the WEF. And so, without further ado, I present to you...

'The Great Termonfeckin Ejaculation'

1. We need climate lockdowns now!

I love pandas. I love koalas. I love chia seeds and I love kale. But I recognise that not everyone is as environmentally sensitive as I am and that's why I propose that people should be forced to do the right thing to prevent climate change whether they want to or not. And, just like with anything that is ultimately better for someone, they'll love their new way of life in no time. Goji berries and activated bee pollen for breakfast, fake meat and kale for lunch and insects on the BBQ for dinner, what's not to like?

Similarly, we need to ban meat from school meals. For the sake of our planet, we need to encourage our children to be ethical little angels and so we should inform them that for every mouthful of meat they eat, a little girl in Africa dies from climate change.

Of course, sometimes kids need to be persuaded in more creative ways, especially those who, for whatever strange reason, quite like the taste of meat. But here's the thing: there is nothing kids love more than farting. It's utterly hilarious and soon has them in stitches. So let them eat mountains of lentils and beans and tell them to enjoy the off-gassing effects as much as they like. And once their giggles have all subsided, take the opportunity to impart the important message that the less cows are raised for slaughter, the less cow farts are emitted into the atmosphere and that the compassionate and groovily fun meal choice which they have just made has positively contributed to a net reduction in methane emissions.

In short, one of the best side-effects of Covid lockdowns was the drastically lower levels of pollution, both on the roads and in the air. From here on in, we need to ration, if not limit completely, the use of private cars and control exactly what people eat, lest our planet explodes from the accumulation of CO_2. Time is of the essence. You know, the idea of a central bank digital currency could be very handy for climate lockdowns...people can have a certain petrol or meat 'allowance'. One steak a month and nothing more, as Mr. Biden himself suggested, as much a poster boy for the neurological effects of a low-meat diet as you could hope to find. Lockdowns are a wonderfully adaptive societal policy and we should make sure not to confine their usage purely to Covid....and so I urge world leaders everywhere to make use of them for saving the planet as a matter of urgency!

2. Please just tell us what to do

Which one of us doesn't suffer from a spot of the old existential angst? For me, this was completely cured by the pandemic. I mean, there is nothing like a government confining us to our homes on pain of arrest with the order to watch Netflix and get takeaways to solve that angst in a jiffy. Such scenarios provide order, meaning and purpose to our lives, where before there was none. Viktor Frankl eat your heart out....in the 21st century, the answers to life's deepest questions arrive ready-made. Just think of the saved therapy bills!

So we need more of this going forward, please. Life is hard enough without having to work out what on earth it all means. And takeaways are delicious anyway.

3. We are all more caring and compassionate than we thought

Do you remember how up until a few years ago we all thought that we were lonelier than ever, more isolated than ever, starved of human company and living in a senseless, capitalist world which had all its priorities upside down?

Oh how Covid has changed everything! Indeed, who among us has not felt, as we self-isolated for a fortnight, or as we looked out the window on empty streets: "We're all in this together.... we are doing this for each other." Oh, Maggie Thatcher, there is indeed such a thing as society! I bet Covid has you turning in your grave.

And this compassion has shown itself to be a transferable skill. Indeed, when I first learnt of the unfolding tragedy in Ukraine, I painted my house yellow & blue so as to show everyone just how much I care about their predicament (which is an awful lot, by the way).

And as soon as I heard that refugees would be coming, I contacted my local authorities to say that I had spare rooms available. And sure only yesterday didn't Olga and her teenage son, Igor, arrive. The only slight snag was that they mentioned they were unvaccinated at which point I rang the pound asking if I could do an exchange but I was told that I'd have to make do. I then suggested to Olga that she might consider getting vaccinated, just the first two jabs, she could leave the three boosters if she wanted, but she responded 'Not over my dead body! We're escaping from totalitarian leaders and their diktats, thank you very much' and words of that kind, to be honest it all got a bit out of hand. Anyway, it became quite clear that they weren't up for it but thankfully they've agreed to stay in their rooms for a few weeks while I set about trying to rehome them.

In any event, my point is that it shouldn't really come as a surprise that we are all generally full of the milk of human kindness these days. I mean, sure don't we all have ethical principles running through our veins? Gone are the eras of discrimination, segregation or of forcing people to do things to their body that they don't want to do. My body, my choice. It's truly no wonder to me that the crimes against humanity that have been perpetrated by the antivaxxers have led to widespread support among the right-thinking among us for mandatory vaccination and for keeping these crackpots as far away from the rest of us as possible. I'd bet that even if we held them down and forcibly injected them, they still wouldn't express even a modicum of gratitude for the manifest health benefits that would then occur. Right pain in the proverbial, the lot of them.

4. The information war is half the fight

We are so lucky to live in this day and age. Imagine if Covid had happened 300 years ago. The outbreak in Wuhan would probably have been reported in *The Times* only two months later and no doubt just in a small news item on the foot of page 7 at that. And sure, by that stage, the virus would already have reached London and all its physicians would have been under the erroneous impression that they were merely witnessing a bad flu year and no doubt treating their patients on that assumption. And so everyone would have been none the wiser and probably gone about their lives as normal, as obvious a case of ignorance not being bliss as you can possibly imagine.

And sure in the back of beyond in the American plain, the Mayan jungles or the Andes, no one would ever even have heard of it at all while the pastoralist tribes of Tanzania would have continued their goat-herding catastrophically ignorant of the fatal danger that they (and their goats) were in.

Now, can you honestly imagine anything more horrific than such a scenario? Such a possibility strikes me as the dystopia of all dystopias, to be utterly frank.

So we should all be grateful that governments and tech companies were able to inform us about everything to do with the virus, the latest fatalities and case numbers, and exactly how we should think about its various shenanigans. I can't remember the name of the luminary who said that there is nothing like a radio in each household to make everyone fall head over heels to do their government's bidding, but he was one intelligent cookie, that's for sure. So, here's to the ever-closer merging of big tech and governments! Hip, hip, hooray!

5. We need to be tougher on anti-vaxxers from the get-go

I don't know about you but I found it utterly horrifying to learn that so many people of a tinfoil hat and conspiratorially minded disposition even existed. To my mind this clearly raises the question of why such people have not been kept under close tabs all along. For example, why is it that governments are happy to keep suspected terrorists under surveillance but not anti-vaxxers? Sure terrorists are only responsible for the deaths of a handful of people every now and again whereas anti-vaxxers are arguably to blame for millions of deaths worldwide and, as such, pose a far greater threat to the national security of any government. So, at minimum, these people need to be monitored alongside other terrorists.

However, I am strongly of the view that governments need to go much further when it comes to the anti-vaxxer problem. In particular, my own major recommendation is that anti-vaxxers should be obligated to wear tinfoil hats while in public so that good and law-abiding citizens will know to keep their distance. This is not just because being in too close proximity might risk exposure to dangerous airborne droplets but also because it might run the risk of picking up their odd views, whether from overhearing their neurotic self-mumblings or simply because their demeanour naturally imparts a certain contrary view of the world. Similarly, I think special camps should be constructed where the willingly unvaccinated can live for the duration of any pandemic and infect each other as much as they like. I've always wondered exactly what was the point of County Offaly, for example...could it not be turned into a large-scale camp for the unvaccinated? In my view, this would be a more than adequate and hopefully final solution to the anti-vaxxer problem.

So, there you go, there are the five principal points of The Great Termonfeckin Ejaculation, all of which need to be acted on urgently, not just because Covid isn't going anywhere but because there already exist new health threats on the immediate horizon. What was it they called it, Monkeypox or something like that? We clearly need to develop a vaccine for this new disease as a top priority and, of course, monkeypox health passes whereby people have to prove not only that they are vaccinated but also that they are not a monkey. Actually, I saw a very sad case of what is likely some new kind of plague the other day, a fella on the outskirts of Termonfeckin, braying away to himself, no doubt suffering from the world's first case of Donkeypox. We live in a very dangerous world, of that we can all be sure and we need to be ready for all eventualities.

But as dangerous as the world is, I am still very optimistic for its future, mainly because of the visionaries who are leading us in the right direction. Indeed, only the other night I had the most beautiful, beautiful dream about our collective future....

Oisín's Dream of the Future: The World in 2030

Look, I know that talking about all the anti-vax nonsense has probably been a bit of a downer at times but I wish to end this book on a really positive note. Indeed, simply put, I believe that there is so much to look forward to. Counting my blessings before someone takes a hatchet to them has always been one of my favourite pastimes and I've been so generally over the moon lately that even my subconscious has been on the case. Indeed, I had the most marvellous dream the other night. It still gives me goosebumps when I think of it. I wish to record it here as a prophetic portrayal of the world in 2030, a portrayal which I hope, pray and believe will come true....

"It is early in the morning and I have just woken up. As always, I love to talk to my dear wife first thing in the morning, there is nothing like whispering sweet nothings and sharing dreams about this and that as dawn's rosy feet rise over the city. And so I immediately open up Zoom and give her a call.

"Assumpta, darling, how's it going up there in the attic?"

"Ah, sure, it's grand, Oisín. You'd think you'd tire of it after a year but actually there is always a cobweb to clean and sure you can see the mountain of jumpers I've knitted behind me."

"And you're not at all bored, are you, dearest?"

"Sure not at all. I mean, we all have to do our bit and if the experts say we all have to stay in separate rooms in order to flatten the curve, then that's just what we have to do, isn't it?"

"That curve will be flattened one day, Assumpta and what a day that will be!"

"Oh, it will be the flattest curve ever, my darling! I can't wait…. ah hang on, it's 8 AM, isn't there an announcement now from our World Leader for Life?"

"Oh, so there is! Well remembered, I'll turn on the radio now."

'And so we now go live to Switzerland where our glorious World Leader for Life is addressing the Celtic Isles Protectorate.

"Good morning, my little sheep. How are you doing today? I have such wonderful news for you all. Indeed, it seems that the current situation with the Omega Plus Plus Plus za.3 subvariant is settling down across the Celtic Isles Protectorate, with the exception of the Galway region. Therefore, it is only a matter of a short period of time, maybe only a few months, before socializing within households will be permitted once again and, indeed, that citizens will be allowed to stand on their front doorstep. If all continues to go well, short walks up to the front gate might be feasible by the early summer.

But my message this morning is not all entirely positive. To those of you who have not taken up the call to receive your 52nd booster, even after our second warning, we have a very important message. Look outside, yes, right now, that's right…. what do you see? Do you see those men standing outside your house in Hazmat suits? Those are your local Covid Protection Officers and they are here to take you away. Goodbye, you, naughty, naughty little sheep who didn't want to be part of the flock anymore, goodbye…"'

The broadcast ends and I hear screaming. I rush over to the window and look outside. "Oh my, Assumpta, it's our neighbours, Séan and Sandra, they are being taken away by a team of CPOs!"

"Oh, how horrifying, Oisín, what a world we live in now...."

"I know, to think that we were living next to antivaxxers all this time!"

"It's too awful to contemplate. But, on the other hand, I feel so much safer now knowing that they are gone."

"Me too. Safety before everything! Say, are you all right for supplies up there, dearest?"

"Oh absolutely, I've got plenty of tinned lentils to keep me going. And I think I'll have kale and vegan eggs for breakfast. I still don't know how they were able to make them, sure you could never tell they aren't from a hen. And there is no cholesterol in them!"

"But is it any surprise really, Assumpta, when you think of what they've been able to do with the vaccines, now successfully modified for the 33rd time to deal with all the variants Covid, the divil that it is, has tried to come up with..."

"You are so right, Oisín. Boy, are we lucky to be alive in these times."

"That we are, that we are, so very, very lucky."

And we that, I awoke, all smiles, a warm glow over my whole body.

Ah, may all of this come to pass and may we all be that lucky!

So join me, dear reader, and do as I have done, and fight.... fight for our future! For is it not true, as I believe was once said in Toy Story, that there is not enough darkness in the whole world to put out the light of one small candle.[18]

---

[18] Was it not Tolstoy who said that, Oisín? (Ed.)

Printed in Great Britain
by Amazon

10101889R00081